Hamlyn all-colour paperbacks

Prue Napier

Monkeys and Apes

illustrated by David Nockels

FOREWORD

Primate biology, which this book is all about, is a subject of special interest to mankind today. Primate biology is the science that deals with man's zoological background, the roots of his humanity and the springboard of his future development. George Gaylord Simpson, the great American zoologist, has said: 'The past of organisms is one of the determinants of their future.' The lives of primates provide the key to the life of man.

Primates are an unique group among the mammals. They are characterized by the possession of two outstanding attributes – a generalized body form and a specialized brain. This combination has reached its peak in man and has enabled him to colonize the world from pole to pole, to explore the ocean depths, the highest mountains of the earth, and the craters of the moon. Man's technological achievements tend to blunt his awareness of his biological heritage and to reinforce his sense of superiority over the natural world. The study of monkeys and apes and their role in human evolution redresses the balance, and reveals man's true place in nature, his oneness with the animal kingdom.

Acknowledgements and thanks are due to the many scientists whose work has been drawn upon extensively in this book. I would like, particularly, to thank Sir W. E. Le Gros Clark, Prof. A. H. Schultz and Dr John Napier whose research has formed the basis of the sections on structure and function. I am also indebted to the following: Dr C. R. Carpenter, Dr Irven De Vore, Dr John Eisenberg, Dr John Ellefson, Mr Philip Hershkovitz, Mr J. E. Hill and Dr W. C. Osman Hill; and also to Mr L. G. Smith, M.B.E., sometime Head Keeper of the Monkey House, London Zoo.

I most gratefully acknowledge the help of the aforementioned and that of all the primate biologists, too numerous to mention, whom I hope will be pleased to find themselves involved in this book, one way or another.   P.H.N.

Published by The Hamlyn Publishing Group Limited
London · New York · Sydney · Toronto
Hamlyn House, Feltham, Middlesex, England
In association with Sun Books Pty Ltd Melbourne

Copyright © The Hamlyn Publishing Group Limited 1970

ISBN 0 600 00075 3
Phototypeset by Filmtype Services Limited, Scarborough
Colour separations by Schwitter Limited Zurich
Printed in Holland by Smeets, Weert

CONTENTS

INTRODUCTION

Monkeys and apes are man's closest living relatives and therefore as a group they are of particular interest to humans. Men, monkeys and apes resemble each other much more closely than they resemble other mammals such as dogs, horses, bats or whales. This similarity is the result of a common inheritance which is expressed zoologically by including all three – with the lemurs, lorises, tarsiers and treeshrews – in a single order, Primates.

Many monkeys and apes look very human, and often seem to behave in a human way. They play, investigate, manipulate new objects, learn fairly quickly and communicate with each other; they sometimes use tools to obtain food and even occasionally make them.

Monkeys and apes form complex social groups and develop behavioural patterns which are often reminiscent of the structure of human societies.

Undoubtedly many people find this similarity disconcerting. Monkeys and apes are at the same time too much like people for the relationship to be ignored and too different for it to be freely acknowledged. Some find the appearance of non-human primates grotesque, or even disgusting, and sub-consciously reject their close relationship. Others tend to attribute human reactions and motives (anthropomorphism) to monkeys and apes which is most misleading as they are not conditioned by human ethics and prejudices. Each species has its own patterns of behaviour.

Moreover, most people are only able to see monkeys and apes under captive conditions, usually in zoos or circuses where they are living abnormal lives in unnatural surroundings. To see them at their best and to appreciate their grace and beauty, it is essential to study them in their natural environment. Although extensive field studies have only been carried out on a few species, they have revealed varied and subtle complexities in the lives of wild primates. These observations have led to many conclusions, some of which have shed new light on aspects of human behaviour that could never have been guessed at by looking at a few captive animals through the bars of a cage.

Howler

Pinché

Ruffed Lemur

Colobus

Potto

Sifaka

What are primates?

Primates are mammals that separated from the primitive mammalian stock some sixty-five million years ago. Mammals differ from the other vertebrates (animals with a vertebral column such as fish, amphibians, reptiles and birds), in having a very advanced system of reproduction. The young are protected within the mother's body before birth and are nourished by the secretions of special milk glands (mammary glands) after birth. In addition mammals are warm-blooded, with fur-clad bodies to insulate them from the cold. Birds are the only other warm-blooded vertebrates.

As a group primates have a complex and advanced brain, compared with other mammals, a well-developed visual system, and a body that is specially adapted for living in trees but which can also function successfully in other habitats. The order Primates includes two suborders, the anthropoids (man-like forms such as the monkeys, great apes and man) and the prosimians. Literally translated, prosimian means pre-monkey or early monkey. This group includes treeshrews, lemurs, lorises, galagos (or bushbabies) and the tarsiers. The order also includes all extinct primates that are known only from fossil remains.

Living primates include certain key species that illustrate the major stages in their evolution, as shown by the study of fossilized skeletons of animals that became extinct millions of years ago. The early primate stage, of about sixty-five million years ago, is represented by the treeshrews; although some zoologists consider them to be more closely related to the insectivores they have many primate characters. The final stage is man, the most successful species of all, at least as far as total numbers and range of distribution is concerned.

Many zoos have specimens of these key primate species so that some of the stages in the evolution of primates can be visualized easily.

The evolutionary progress of the group is represented by the following living primates in ascending order of evolutionary improvement: the treeshrews, the lemurs, the tarsiers, the New World monkeys of South and Central America, the Old World monkeys of Africa and Asia, the apes and finally man.

Man

Chimpanzee

Gibbon

Bushbaby

Indris

Aye-aye

Primate families
Prosimians

In classifying the primates, as with all other groups of animals, zoologists divide them into families; all the members of a family have certain features in common. The two suborders of the primates, the prosimians and the anthropoids, are each divided into six families. Each family of living primates will now be briefly mentioned.

Prosimians are found in Africa, Madagascar, and South East Asia.

Treeshrews (family Tupaiidae) are thought to be rather similar to the primitive mammalian stock from which the primates evolved. Treeshrews are the least monkey-like of all the primates; they have long snouts, and claws on both hands and feet. They live in the forests of India, Burma, south-western China and South East Asia.

Lemurs (family Lemuridae) include the true lemurs, dwarf lemurs, and mouse lemurs. The term 'lemur' is sometimes used to include the other Madagascan primate families but here it is being used in its more restricted and correct sense. Lemurs

Tarsier

Loris

Ring-tailed Lemur

Feather-tailed Treeshrew

Common Treeshrew

Mouse Lemur

have flat nails on both hands and feet. They are only found on Madagascar and some nearby islands.

Indrises (family Indriidae) are similar to lemurs but with longer legs. The family consists of the sifakas, Indris, and the Avahi or Woolly Lemur. They are also only found on Madagascar.

The Aye-aye (family Daubentoniidae) is placed in a separate family on account of its rodent-like teeth and peculiar wire-like middle finger. Aye-ayes are nocturnal animals and there are only a few remaining in the forests of Madagascar.

Lorises (family Lorisidae) are also nocturnal animals; they include the slow-creeping lorises and Pottos and the agile leaping galagos or bushbabies. Like the lemurs their fingers and toes bear flat nails. Galagos and Pottos are found in Africa; the Slow and Slender Lorises are found in the forests of South East Asia, India and Ceylon.

Tarsiers (family Tarsiidae) are small nocturnal animals, characterized by their peculiar leaping locomotion and enormous, forwardly facing eyes. They live in the forests of Sumatra, Borneo, the Philippines and Celebes.

Orang-utan *(above)*,
Lar Gibbon *(below)*

Anthropoids

The anthropoids (monkeys, apes and man) have a world wide distribution. There are six families which are placed in three major groups.

The first group, the New World monkeys, includes the marmosets and the monkeys of South America.

Marmosets (family Callitrichidae) include the true marmosets, Goeldi's Marmoset and the tamarins. They have claws on both hands and feet and a flat nail on the big toe. They live in the forests of Central and South America.

Cebids (family Cebidae) include spider monkeys, woolly monkeys, howlers and capuchins (with prehensile tails) and squirrel monkeys, the Douroucouli, titis, sakis and uakaris (without prehensile tails). All are found in the forests of Central and South America. They have pointed, rather curved nails on the hands and feet.

The second major group of the anthropoids contains all the Old World monkeys which are placed in one family.

Old World monkeys (family Cercopithecidae) are divided into two subfamilies: one includes the omnivorous guenons, mangabeys, baboons and macaques; the other, the leaf-eating langurs and colobus monkeys. Most Old World monkeys have long tails, none of which is prehensile; some have short tails and others have only a minute stump. They live in the forests and grasslands of Africa, India, South East Asia, China and Japan.

The apes and man together form the third major division of the anthropoids.

Gibbons (family Hylobatidae) include the gibbons and the Siamang. They are tailless and have extremely long arms. They are widely distributed throughout the tropical forests of South East Asia.

The great apes (family Pongidae) include the Orang-utan, the Gorilla and the chimpanzees. Gorillas and chimpanzees are found in the tropical forests and woodlands of Africa, the Orang-utan is confined to Sumatra and Borneo.

Man (family Hominidae). Both living and extinct forms of man are placed in this family, the only living representative of which is modern man (*Homo sapiens*). Extinct forms include Neanderthal man from Europe, Java man and Pekin man from the Far East and the australopithecines from Africa.

Capuchin *(top)*, Diana Monkey *(centre)*, marmoset *(bottom)*

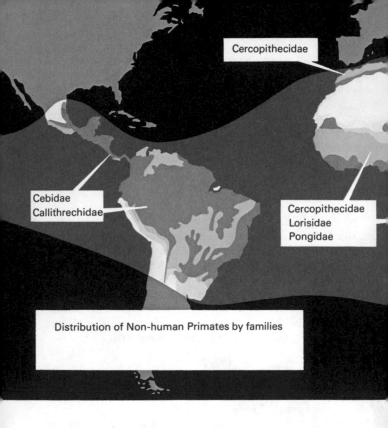

Cercopithecidae

Cebidae
Callithrechidae

Cercopithecidae
Lorisidae
Pongidae

Distribution of Non-human Primates by families

WHERE PRIMATES LIVE

All non-human primates are inhabitants of the warmer areas of the world. Unlike some mammals, such as squirrels, non-human primates have never evolved the adaptation of hoarding food gathered when plentiful. Most of their food of fruit and leaves is perishable, thus they are dependent on finding sufficient food every day of the year to meet their needs.

Life in tropical forests is relatively easy for most primates. On the Equator, all the year round, the daylight lasts about twelve hours out of the twenty-four, so there is ample time for foraging and feeding; the supply of fruit, leaves, flowers and insects is plentiful and not seasonal in any way. In the temperate zones of the world, life is much harder and the onset

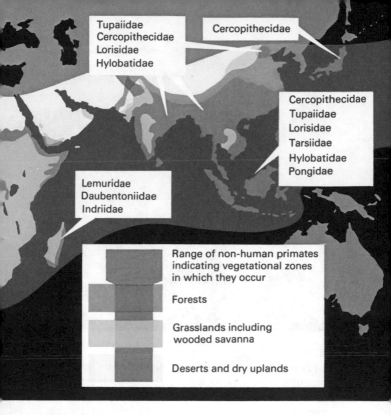

Tupaiidae
Cercopithecidae
Lorisidae
Hylobatidae

Cercopithecidae

Cercopithecidae
Tupaiidae
Lorisidae
Tarsiidae
Hylobatidae
Pongidae

Lemuridae
Daubentoniidae
Indriidae

Range of non-human primates indicating vegetational zones in which they occur

Forests

Grasslands including wooded savanna

Deserts and dry uplands

of winter brings not only a reduction in the length of daylight but also a scarcity in the food supply. Here there is a limited growing, flowering and fruiting season, and in winter most of the trees are leafless. In winter, therefore, there is less food and less time in which to search for it.

Low temperatures seem to be a less important factor in limiting the spread of primates into high latitudes, than the amount of daylight during the winter. The highest latitude inhabited by a non-human primate is 41° 20′ north, on the Japanese island of Honshu. In December and January the daylight lasts for only nine hours, the midwinter snows cover the ground and the monkeys are forced to subsist on bark which they strip from the branches of the leafless tress.

In the trees

Most monkeys and apes live in forests; some live on the forest floor but the vast majority are found among the leaves and branches of the forest canopy, where predators are few and abundant plant and insect food is available all the year round.

Monkeys and apes depend for their survival upon their skill and agility in the trees, the result of their exceptional co-ordination of eyes, brain and limbs which permits them to run, leap and swing among the branches in perfect safety high above the forest floor.

Man usually sees tropical forests from an aeroplane as a brilliant unbroken carpet of green far below. But on the forest floor twilight prevails, as the dense foliage of the canopy shuts out the rays of the bright equatorial sun.

Primates which live mainly on the forest floor are usually heavily built forms, such as the Mandrill and the Gorilla. The Gorilla, for instance, because of its vast size, needs large quantities of plant food, such as wild celery which only grows on the ground. Furthermore, male Gorillas, which may weigh

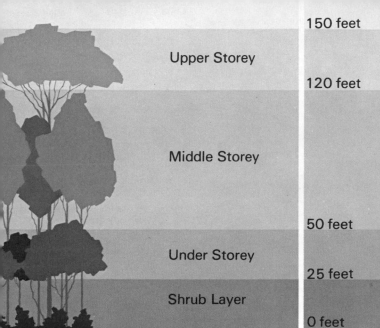

	150 feet
Upper Storey	
	120 feet
Middle Storey	
	50 feet
Under Storey	
	25 feet
Shrub Layer	
	0 feet

Vertical section through a tropical rain forest

up to 450 pounds (203 kg.), find all but the stouter branches inadequate to support their weight and build their sleeping nests on the ground. The more lightly built females and juveniles can climb into the trees much more freely.

The upper storey of the canopy, where the crowns of the trees are umbrella-shaped and separated by wide gaps, is usually the home of the more acrobatic monkeys. The distribution of fruit and leaves at the edges of the crowns of these emergent trees requires that the monkeys should be extremely agile and capable of balancing on narrow flexible supports, which can be as much as 150 feet (50 metres) above the ground. Moreover, the gaps between the crowns mean that monkeys moving through the tree-tops must often take tremendous leaps.

Monkeys living in the lower levels, where the crowns of the trees are more tightly packed and the situation is thus less hazardous, do not need to be so agile or have such great powers of leaping to enable them to find their food and escape from their predators.

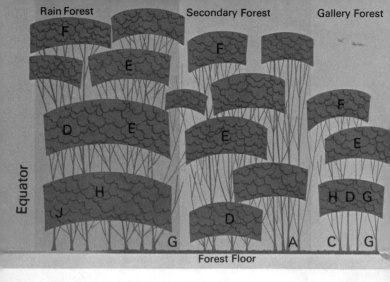

Rain Forest   Secondary Forest   Gallery Forest

Equator

Forest Floor

On the ground

The diagram shows the effect of decreasing annual rainfall on the vegetation of the northern part of Africa, from the River Congo on the Equator northwards to the Tibesti plateau on the Tropic of Cancer; a distance of about 1,300 miles (2,380 kilometres). The tropical rain forests and swamp forests cover a considerable distance on each side of the Equator where the rainfall is plentiful and frequent. Travelling northwards the rainfall gradually diminishes and the rain forest is replaced by wooded grassland, called savanna. This is characterized first by patches of tall grasses amongst the trees which gives way to small groups of trees in open grassland. Occasional strips of tropical forest are found along the banks of rivers, where there is sufficient moisture to support them; these are the so-called gallery or riverine forests.

Further north still conditions are much drier, and the vegetation is subjected to frequent fires. The only trees that can survive in this steppe or semi-desert zone are the tough acacias and thorn scrub. Finally, the desert proper is reached where the rainfall is practically nil and in places there is no plant life at all.

The tropical forests of today are not as extensive as they once were. At the time of the earliest primates (about sixty-five

Habitats of some African primates

A Olive Baboon
B Patas Monkey
C Grivet Monkey
D Mona Monkey
E Greater White-nosed Guenon

F Red Colobus
G Sooty Mangabey
H De Brazza's Monkey
J Allen's Swamp Monkey
K Moustached Monkey

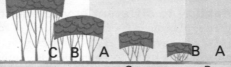

oded Savanna   Savanna   Desert

Tropic of Cancer

million years ago), many of the present savanna and semi-desert areas were covered with forest as the climate of the whole area was much wetter. As the forest areas dwindled, certain groups of primates deserted the rain forests to live on the ground in the newly formed savannas. Two of the groups that left the forest trees some twenty million years ago are thought to have been the ancestors of the baboons and the ancestors of man.

Baboons that live in the savanna, steppe and sub-desert zones of Africa spend much of the daylight hours foraging on the ground and have a very varied diet. They pluck grasses and seeds; they dig in the ground for tubers and roots; they overturn stones to find insects; they occasionally kill and eat small mammals such as hares and newborn gazelles. But however ground-adapted they have become baboons still seek the protection of trees at night. Before dusk they climb high up into the branches or, in the drier treeless zones, they seek steep rocky outcrops and inaccessible cliffs to avoid nocturnal predators, and remain there until after dawn. Even during the day, baboons must be ready to climb quickly into the slender branches of the nearest available tree to avoid one of their most dangerous predators – the Lion.

Rock of Gibraltar

Adaptability to different environments

Some groups of primate species are capable of adapting themselves to life in a wide variety of environments. For example macaque monkeys are a large and successful group which can live in a variety of habitats in different climatic regions.

The Barbary Ape (*Macaca sylvanus*) lives in the mountains of Morocco and Algeria, and there is a small population on the Rock of Gibraltar. It is the only free-living monkey in Europe but it was probably introduced there several centuries ago. The Gibraltar monkeys are semi-tame and depend on man for most of their food. Those living in Morocco and Algeria, however, are completely wild. This species is able to survive fairly rugged conditions.

Crab-eating Macaques (*M. fascicularis*) inhabit the mangrove swamps and muddy tidal creeks of the mainland and tropical islands of South East Asia. The temperature and the amount of daylight vary very little throughout the year. As their name

Mangrove swamp

Indian town

suggests, Crab-eating Macaques eat crabs and other shellfish which abound in this habitat. They are excellent swimmers.

Rhesus Macaques (*M. mulatta*) live in India, southern China, Thailand and Vietnam. In India they are given a semi-sacred status by the Hindus, and are therefore to some extent protected. They inhabit temples where they are fed by the worshippers. Rhesus Macaques also inhabit towns and villages where they can raid crops and bazaars with impunity.

The most northerly species of monkey is the Japanese Macaque (*M. fuscata*). They inhabit the temperate forests of Honshu and other islands. During the winter months the mountains where they live are covered with snow. The monkeys seek sheltered areas and to keep warm they huddle together and bathe in the warm springs.

These species illustrate the range of adaptability within a single genus of monkey. Most primates, however, are tropical forest dwellers.

Honshu Island, Japan, in winter

EVOLUTION OF THE PRIMATES

Vertebrate calendar

The evolution of the vertebrates can be traced over many millions of years through their fossil remains. Prior to 600 million years ago, only simple soft-bodied forms of life existed which left almost no trace so very little is known about them. Animals with protective shells or hard bones evolved later, some of which have been preserved as fossils. Comparison of earlier and later fossils shows that animals developed gradually from simple to more complex forms, from animals that could live only in water such as the fishes, to air-breathing amphibians, reptiles, birds and mammals.

Equating this 600 million year span of life with a calendar year, primate evolution is a relatively recent event, and modern man is a last-minute arrival on the vertebrate scene.

Significant datelines

23rd March	**Earliest vertebrates.**
21st April	**Primitive jawless fishes** appeared.
15th May	**Bony fishes** and sharks evolved; also the lobe-finned fishes, which are considered to be the fore-runners of the amphibians.
1st June	**Amphibians.** They were the first vertebrates to invade the land. Amphibians lived in swamplands or at the edges of lakes, for they must return to the water to lay their eggs.
3rd July	**Early reptiles** were the first vertebrates able to live the whole of their lives on land. Reptiles are cold-blooded; their body temperature is always the same as that of their environment.
28th July	**Spread of reptiles.** They diversfied and spread, dominating the other forms of life.
25th September	**Primitive mammals** evolved. These rat-sized creatures, the first warm-blooded animals, were clad in fur which insulated them from environmental changes, permitting them to be active all the year round.
September and October	**Heyday of the giant reptiles.** At the end of this long period the dinosaurs became extinct. The mammals were small and relatively insignificant.
21st November	**Spread of mammals.** The mammals filled the vacuum left by the vanished giant reptiles becoming the dominant form of life on land.

27th November	**Earliest true primates.** Treeshrew-like primates (*Plesiadapis*) appeared in North America.
12th December	*Aegyptopithecus,* the first ancestral ape evolved.
22nd December	Earliest ancestors of man (*Ramapithecus*).
31st December	At 8.15 p.m. modern man (*Homo sapiens*) stepped over the threshold.

Vertebrate Calendar

January

Invertebrates only

February

March

Earliest Vertebrates

April

Jawless Fishes

May

Boney Fishes

June

Amphibians

July

Earliest Reptiles

August

Spread of Reptiles

September

Primitive Mammals

Giant Reptiles

October

Primitive Mammals

Giant Reptiles

November

End of Giant Reptiles

Spread of Mammals

Earliest Primates

December

Earliest Ancestor of Man

Homo sapiens

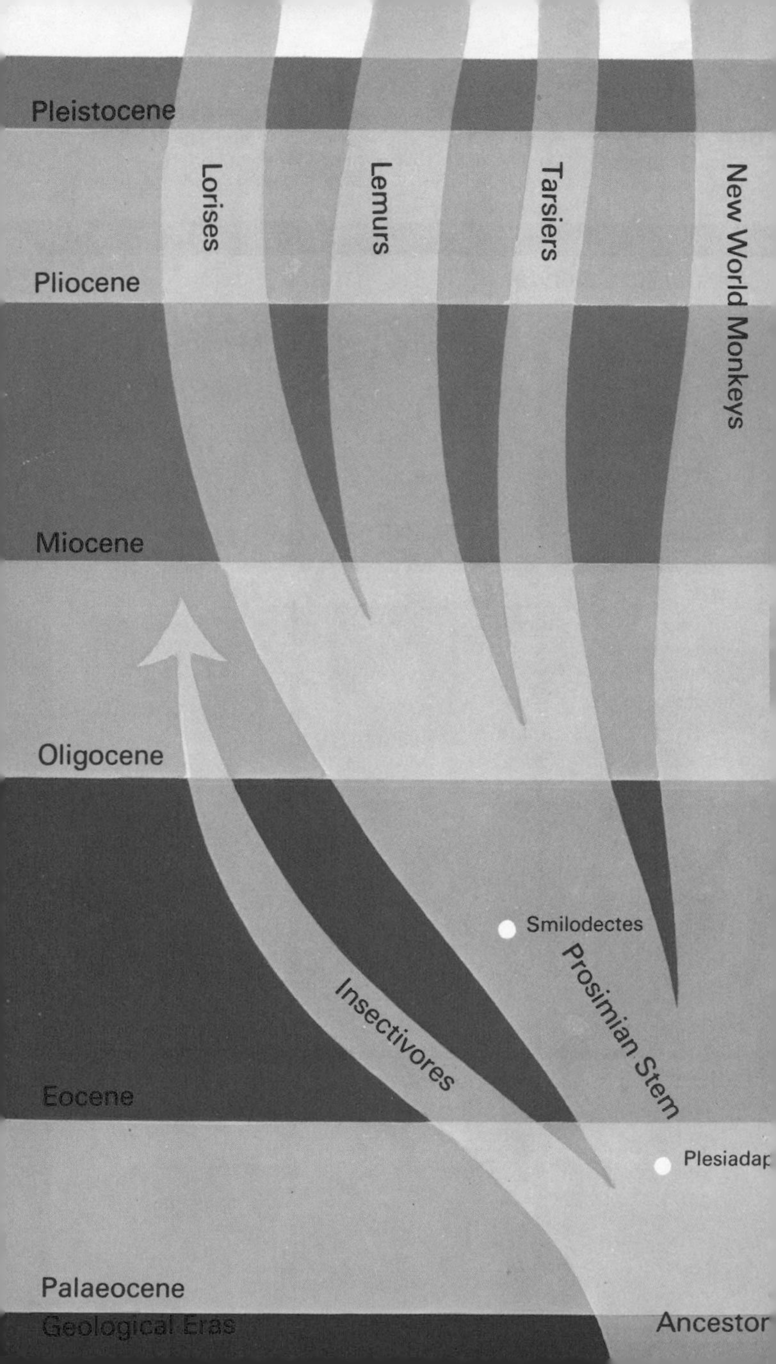

Pleistocene

Lorises · Lemurs · Tarsiers · New World Monkeys

Pliocene

Miocene

Oligocene

Smilodectes

Insectivores · Prosimian Stem

Eocene

Plesiadap

Palaeocene

Geological Eras · Ancestor

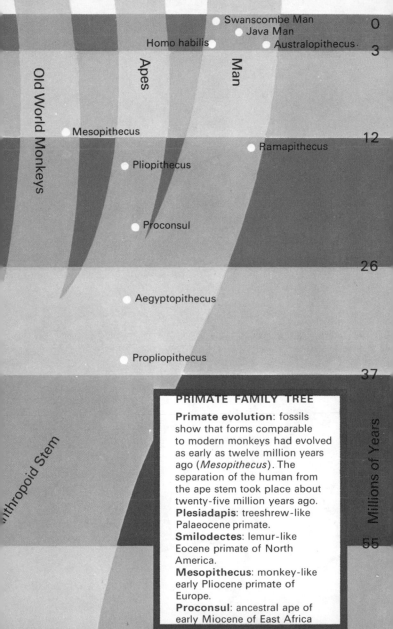

Old World Monkeys

Apes

Man

0
Swanscombe Man
Java Man
3
Homo habilis
Australopithecus

12
Mesopithecus
Ramapithecus

Pliopithecus

Proconsul

26

Aegyptopithecus

Propliopithecus

37

Anthropoid Stem

55

Millions of Years

65
Insectivores

PRIMATE FAMILY TREE

Primate evolution: fossils show that forms comparable to modern monkeys had evolved as early as twelve million years ago (*Mesopithecus*). The separation of the human from the ape stem took place about twenty-five million years ago.
Plesiadapis: treeshrew-like Palaeocene primate.
Smilodectes: lemur-like Eocene primate of North America.
Mesopithecus: monkey-like early Pliocene primate of Europe.
Proconsul: ancestral ape of early Miocene of East Africa

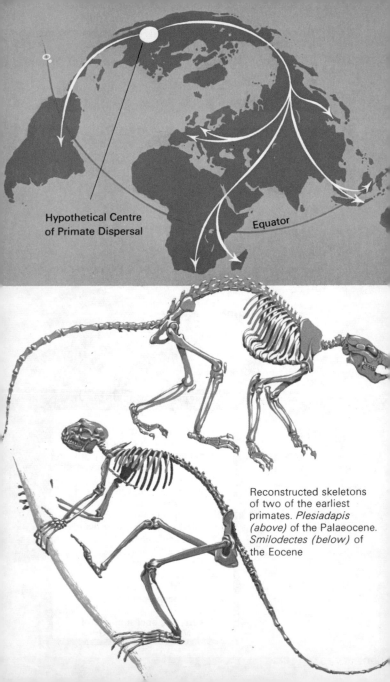

Hypothetical Centre
of Primate Dispersal

Equator

Reconstructed skeletons
of two of the earliest
primates. *Plesiadapis
(above)* of the Palaeocene.
Smilodectes (below) of
the Eocene

Earliest primates

The basic mammalian stock consisted of small rat-sized creatures, usually classified as insectivores. In the early stages of the Palaeocene epoch, about sixty-five million years ago, primates became distinct as a separate group of mammals.

Plesiadapis is the earliest known primate and has been found in both Europe and North America. From the form of the fossil skeleton it is thought that these creatures lived on the forest floor.

At this time, in the Palaeocene, primates were small animals with comparatively long bodies, long tails and short limbs. Their snouts were long, indicating their great reliance on the sense of smell: their eyes were small and faced sideways, showing that they possessed little binocular (or three-dimensional) vision. All five digits of both hands and feet bore claws. Many millions of years were to pass with many intermediate stages, before the primates evolved into the characteristic forms of prosimians, monkeys, apes and men that we know today.

The earliest primates were forest-dwelling creatures, and from their 'birthplace' in the northern hemisphere, they migrated freely between the Eurasian and North American continents, via the forested Bering Strait land bridge which then united north-east Siberia and Alaska. In the early Eocene, about fifty-five million years ago, the climate of the Bering land bridge (latitude 55° north) became colder and the forests gradually disappeared, putting an end to this free interchange. From then on the primates evolved separately in Eurasia on the one hand and in America on the other.

Smilodectes is an example of an early fossil prosimian. Fossils have been found in the middle Eocene (forty-five million years ago) of North America. The skeleton indicates that it was a tree-living animal, somewhat similar to the modern sifakas of Madagascar.

Eventually the Eurasian primates invaded the whole of Africa, Madagascar, southern Asia and the East Indies. There they evolved into the prosimians, Old World monkeys and apes. The North American stock migrated over the isthmus of Panama into South America, where they gave rise to the present-day New World monkeys and marmosets.

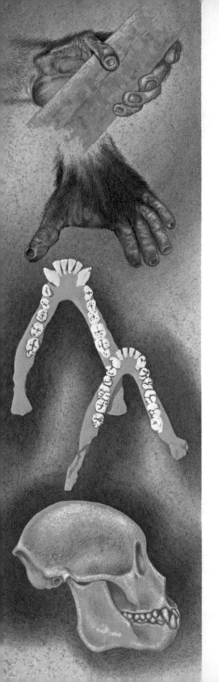

Arboreal adaptations

When primates first evolved from the basic mammalian stock, they occupied some of the numerous niches of the forest floor. They were to be found nesting and scurrying about among fallen logs, and roots of trees, even perhaps in underground burrows. For food they would have foraged among the debris of the floor, and in the shrubs and low trees.

By the middle Eocene, primates were already arboreal creatures with large front-facing eyes and rather short noses (indicative of a greater reliance on vision than on smell). Their legs were much longer than their arms and their big toes, widely separated from their other toes, gave the foot a grasping or prehensile function; this type of toe is referred to as 'opposable'. The earliest tree-living primates adopted a vertical clinging posture on trunks of

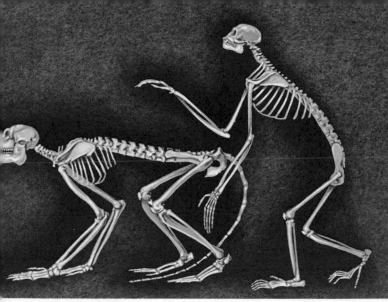

trees or vertical branches while at rest; their long legs were acutely bent and held close to the body. Movement was achieved by a powerful straightening action of both legs, which propelled the animal from branch to branch. Loco-motion was therefore a form of tree-hopping.

By the late Eocene, these early monkey fore-runners had evolved opposable thumbs which enabled them to grasp the slender branches of the high forest canopy. The tips of their fingers and toes were broader and, instead of claws, were sur-mounted by flat nails; their hands had become not only pre-hensile but important sensory organs from which much valu-able information about the environment could be transmitted to the already expanding brain. Subsequent changes in bodily form were not great. The most characteristic trend, already apparent in the Miocene about twenty million years ago, was the lengthening of the arms relative to the legs; this led ultimately to the arm-swinging of gibbons and spider monkeys.

The hands and feet of the majority of primates are capable of grasping and have flat nails on the ends of the fingers and toes. Each extremity has one large and extremely mobile digit, the thumb or the big toe.

Semi-bipedal savanna-dwelling Miocene primates

The beginnings of ground-living life

In the early stages of their evolution, primates were essentially forest-dwelling animals, some living on the ground and some in the trees. In fact they had few alternatives, for much of the temperate and tropical regions of the earth was covered with forests. Here they evolved many of the characters by which we recognize primates today – highly mobile limbs, prehensile hands and feet, long tails, frontally facing eyes and short faces. All these physical characters and many aspects of primate behaviour are adaptations to life in the trees.

During the Miocene the climate and therefore the vegetation of the world underwent a profound change. Many forests were replaced by grasslands – a relatively new form of vegetation. This new type of habitat was rapidly populated by some of the most progressive and 'adventurous' of all mammals, the primates. It is believed that at this time, about twenty million years ago, two important groups of primates came into existence, one ancestral to the baboons and one ancestral to man.

The invasion of open woodlands and grasslands was the

critical turning-point in the evolution of man. In this environment man acquired the characters that distinguish him from other primates. Although there is no sharp line to be drawn between the bodily form of man and apes, differences do exist in the proportion of certain parts of the body. Man's arms and legs are both relatively long whereas the chimpanzee has exceptionally long arms but fairly short legs. Man's thumb is long, strong, supple and broad; ape thumbs are slender and rather short. Man's brain is much larger in proportion to his body size than is the ape brain. These proportional differences underlie the three main functional characters by which man is distinguished from the apes: man habitually walks upright on two legs; secondly, man possesses a hand capable of fine manipulation; thirdly, man has a large brain and the capacity (through speech) for communicating experience, knowledge and ideas.

The change in habitat, from forest to 'open' country, provided a challenge to certain progressive Miocene primates. Paradoxically, the original adaptations for life in the trees equipped them for life on the ground.

Australopithecus was bipedal and a carnivore

Olduvai Gorge, Tanzania

Earliest true man

At Olduvai Gorge, Tanzania, in 1959 an expedition led by Dr. Louis Leakey, the eminent palaeontologist, made an exciting discovery. He found a man-like jaw of vast proportions in the geological layer known as Bed 1. Scientifically named *Zinjanthropus*, the owner of the jaw is informally known as 'Nutcracker Man'. Stone tools were found associated with 'Zinj' suggesting that the owner was the first tool-maker. In the following year, further discoveries at a slightly lower (and therefore older) level indicated the presence of an even more advanced fossil man, believed by a number of scientists to be the first representative of our own genus – *Homo*. This second discovery was given the scientific name of *Homo habilis*. Among the fossil remains of *Homo habilis* were parts of the skeleton of a hand and an almost complete foot. From these bones, scientists have been able to deduce that this early human ancestor was able to make and use stone tools and, most significantly of all, to walk upright on two feet, almost exactly as modern man does.

The discovery of *Homo habilis* has led to considerable

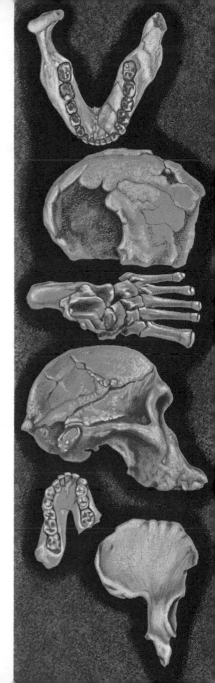

(From top to bottom)
Zinjanthropus jaw; skull and foot
of *Homo habilis*; skull of
Australopithecus; jaw of *H.
habilis*; pelvis of
Australopithecus

controversy. Some experts
on the australopithecines, a
group of 'near-men' from
South Africa first discovered
in 1924, do not regard *Homo
habilis* as the first 'true man'
but as just another member
of the australopithecines.
They point out that the teeth
of *Homo habilis* differ little, if
at all, from those of the australopithecines and the brain
size (as estimated from skull
measurements) was insignifi-
cantly bigger. They also point
out that the australopithecines were equally upright
and bipedal in their walking
gait. The protagonists of
Homo habilis, however, insist
that the minor differences are
significant and indicate that
Homo habilis had crossed the
threshold between 'near-man'
and 'true man'. In support of
the evidence shown by the
fossils, the pro-*habilis* group
point out that at Olduvai
Gorge the remains of man are
actually associated with stone
tools. This critical association
has not been satisfactorily
proved for the australopithecines.

Evolution of modern man

Most anthropologists agree that human evolution in the Pleistocene age, which began about three million years ago, consisted of three main stages or grades.

Firstly, there was the australopithecine stage which includes *Homo habilis* who, notwithstanding the claims made on his behalf to be the first 'true man', is considered to be an advanced member of the australopithecine stock. Secondly, there was the *Homo erectus* stage which included Java man and Pekin man. Fossils of this widely distributed species have been found in Africa, Europe, China and the East Indies. The use of fire was probably discovered by these early men; fire-sharpened spears were used to bring down the swift game animals. They were alive from about 700,000 to 400,000 years ago.

The third and final stage is *Homo sapiens* which includes

Reconstructed skulls *(left to right)* Pekin man, Steinheim man, Neanderthal man, Cro-Magnon man

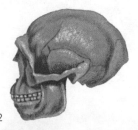

Early man hunting with
fire-sharpened spears

Steinheim man, Swanscombe man, Neanderthal man, Cro-Magnon man and modern man. Swanscombe man and Steinheim man, from England and Germany respectively, are the first known representatives of *Homo sapiens*. They lived between about 400,000 and 100,000 years ago. Neanderthal man, now extinct, but formerly found over much of Europe is regarded as a separate sub-species of *Homo sapiens*, the outcome of a long period of genetic isolation during the last of the Pleistocene Ice Ages, which lasted from 100,000 to 30,000 years ago. Cro-Magnon man is considered to be the direct ancestor of all races and regional varieties of modern man.

These evolutionary stages were, however, structurally continuous; this is shown by estimates of the brain size of successive populations. The brain size of *Australopithecus* was about 435 to 680 cubic centimetres, Java man 775 to 900 c.c., Pekin man 900 to 1,250 c.c. and *Homo sapiens* from 900 to 2,000 c.c.

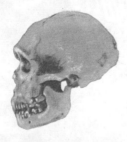

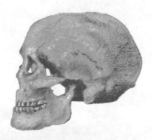

PHYSICAL STRUCTURE OF MONKEYS AND APES

The skeleton

It is difficult to define the primates by listing the characters that set them apart from other orders of mammals. For instance, bats have wings; elephants have trunks; rodents have chisel-like front teeth; and whales have tails and fins for swimming; but primates have no universal 'trade-mark'. Primates can best be described as 'generalized' mammals – that is to say, they are not limited to one particular habitat by specialized physical adaptations such as wings, trunks or fins.

Primates, during their evolutionary past, have lived mainly in trees. Over many millions of years of evolution this challenging arboreal environment has selected and developed certain specifically primate characters such as prehensile hands and feet for grasping, long tails for balancing, large forward-facing eyes for judging distances accurately and long limbs for leaping and climbing. These are all adaptations for moving efficiently through the trees, but do not limit primates to an arboreal life. The langur, whose skeleton is shown below, has basically the same components as that of all prosimians, monkeys and apes. All the adaptations to arboreal life are present but it may spend much of the time on the ground, where it walks and runs easily and gracefully.

Langur skeleton

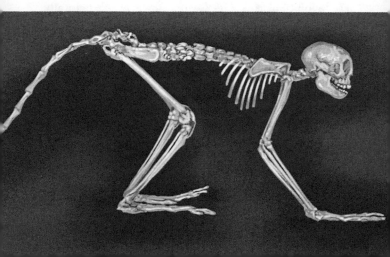

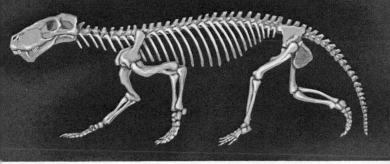

Reconstruction of *Lycaenops* skeleton

The primate skeleton is very primitive and is similar, in arrangement and number of bones, to that of the 230 million years old mammal-like reptile *Lycaenops* of the Triassic age. The limb skeleton which consists of a single bone in the upper arm and thigh, two bones in the forearm and leg and five digits in both hands and feet exactly follows the primitive pattern found among these early reptiles. The differences that exist in living primates are principally those of robustness and proportion.

The table at the foot of the page points out some of the main differences in the skeletons of various types of primates. The most striking differences occur in the size of the brain, and in the presence or absence of a tail.

Main Differences in Primate Skeletons

	Treeshrew	Lemur	Langur	Gorilla	Man
Braincase in ccs	3	18	80	530	1,400
Teeth	38	36	32	32	32
Cervical and Lumbar Vertebrae	22	22	22	22	24
Ribs (pairs)	13	12	12	13	12
Tail	√	√	√	×	×
Hand Bones	27	28	29	27	27
Foot Bones	26	26	26	26	26

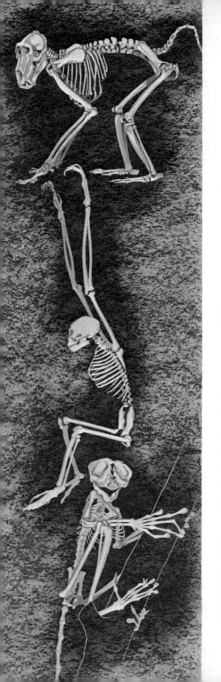

Special adaptations of the skeleton

While many monkeys, such as the langur, have retained a very 'generalized' skeleton, certain primates have become adapted to a particular environment. For example the tiny tarsier lives among slender saplings where it leaps, frog-like, from one upright support to another. Its skeleton is adapted to the stresses of rapid leaping by having, among other things, the lower halves of the two leg bones (the tibia and fibula) fused together and a very elongated heel region. This is the only instance in the whole of the primate order where the long bones of the leg are united.

The primate skeleton varies very little even among animals as different as a baboon, a gibbon, a Gorilla and man; the differences are mainly those of proportion. Ground-living monkeys, such as baboons, have a rather dog-like skeleton with long arms and legs, of almost equal length; they can run and walk on the

Baboons *(top)* are adapted to living on the ground. Gibbons *(centre)* are almost entirely arboreal and swing by their long arms. Tarsiers *(bottom)* have very elongated ankles, an adaptation for leaping.

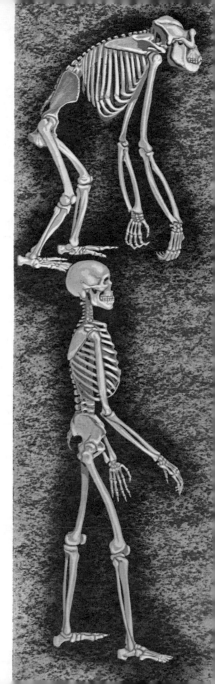

ground as efficiently as other ground-living quadrupeds.

The lightly-built arboreal gibbons have developed long, powerful arms and elongated hook-like hands for swinging beneath the branches of trees; the legs are very rarely used during locomotion and are much shorter than the arms; this type of locomotion is called brachiation. The Gorilla, Orang-utan and chimpanzee also have long arms and short legs, but are much heavier animals; they rarely swing through the trees in this way. The Gorilla and chimpanzee move on the ground in a semi-erect posture, walking on their knuckles, while Orangs progress slowly through the trees using all four limbs.

Man has relatively long legs, an adaptation to bipedal walking, but his arms are also quite long, reflecting his ancestry from an arboreal arm-swinging form, but, it must be stressed, not from a form as specialized as the living chimpanzee, Orang-utan or gibbon.

Gorilla *(top)* knuckle-walking Man *(below)* is bipedal. In both these ground living apes the arms are much longer than those of baboons.

Eyes and noses

During their evolution, primates have gradually become less dependent on their sense of smell and more dependent on their sense of sight. The earliest primates were probably rather like the modern treeshrew which has a long muzzle, with eyes facing sideways. Each eye sees a different image; there is only a small area of visual overlap where both eyes can see the same object. In monkeys and apes, the eyes look directly forward and the face is generally short. Full binocular vision enables them to judge distances accurately, and to appreciate the shape and size of distant objects.

Large eyes enable nocturnal primates to make use of the moonlight or starlight filtering through the trees. Large mobile ears enable them to pick up the faint sounds of the insects and lizards which are their prey.

Treeshrews, lemurs, and lorises all have a naked moist nose, rather similar to that of a dog. A distinct groove runs down the centre of the nose and continues on to the upper lip, which is attached to the underlying gum. All other primates, including the tarsiers, have a dry external nose with no central groove. The upper lip is mobile, not tethered to the gum. This mobility allows a great increase in the range of facial expression and vocalization, which is of great value in communication.

The long muzzles and naked moist noses of the prosimians are indicative of the importance, to them, of the sense of smell. They rely more on their noses to gather information about their environment than do the anthropoids which rely more on their eyes.

Nose shape has been used to distinguish between New and Old World monkeys. In New World monkeys (sometimes called platyrrhines or wide-nosed forms) the nostrils are roughly circular in shape, widely separated, facing sideways. In Old World monkeys (catarrhines or narrow-nosed monkeys) the nostrils are set close together, only separated by a thin septum. However this character varies considerably in different species.

Skulls of treeshrew *(left)*, lemur *(centre)*, langur *(right)*
Heads of tarsier *(top)*, lemur *(centre left)*, chimpanzee *(centre right)*
New World monkey *(bottom left)*, Old World monkey *(bottom right)*

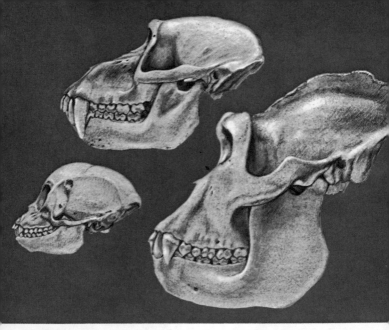

Skulls of squirrel monkey *(left)*, baboon *(centre)*, Gorilla *(right)*

Skulls

All primate skulls are characterized by having the orbit, or eye socket, surrounded by a ring of bone. In non-human primates the relative size of the brain decreases as the body size increases; but the relative size of the jaws increases with body size.

Small monkeys eat insects and fruits; with this type of diet prolonged chewing is not necessary. Larger primates, such as baboons and Gorillas, feed mainly on plants which necessitates prolonged chewing of large quantities of fibrous food. Gorillas spend most of the day feeding, so their jaws and jaw muscles are large and heavy; for the attachment of these big muscles, extra flanges of bone are sometimes present on the top and back of the skull, particularly in large males.

All primates follow the basic mammalian dentition in having incisors, canines, premolars and molars. From the treeshrews to man the number of teeth gradually reduces but there are no significant changes in function.

Lower Jaw	Upper Jaw

Treeshrews have more teeth than other primates, thirty-eight, including three incisors (green) on each side of the lower jaw, but only two in the upper. The two central lower incisors are horizontal and form a dental comb used in grooming.

Lemurs have thirty-six teeth, there are only two incisors on each side of the lower jaw. These teeth together with the canines form a dental comb similar to that of the treeshrew

Squirrel monkeys have small jaws and sharply pointed teeth which can crush the hard bodies of insects. They have the same number of teeth as lemurs but the incisors are upright

Baboons have elongated jaws with thirty-two teeth. They have only two premolars (blue). The canines (orange) are large in males. To accommodate the lower ones there is a wide gap — the diastema — between the upper canines and incisors. The lower molars (red) have four points (cusps) each.

Gorillas also have thirty-two teeth. The canines are more conical and less sharp than those of baboons, but a diastema is present and the premolars are similar. The lower molars have five cusps.

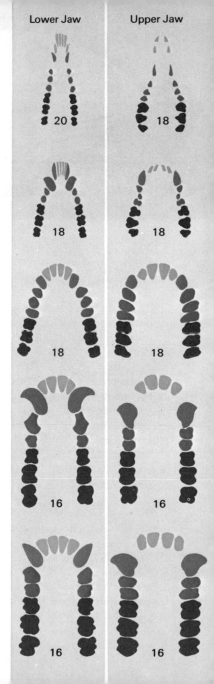

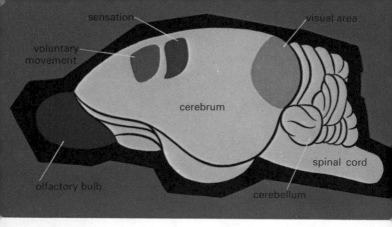

Brains

The brains of primates are very similar to those of other mammals; however, their brains are relatively larger and more complex, which enables them to appreciate and respond to the environment in a more original, less automatic way.

The increase in size has mainly involved the cerebrum which includes the centres for receiving and synthesizing sensations including vision, and for initiating voluntary or conscious movement. The cerebellum, the centre for co-ordination and balance, has increased relatively less. The increase in size has been accompanied by the buckling of the outer covering of the cerebrum into fissures and folds. The folding is most noticeable in the higher primates such as the apes and man.

Primate brains also show an increase in structural complexity; for example the cerebral cortex shows a progressive elaboration of cellular structure by which functional areas (e.g. the visual cortex) can be delineated. These areas are also to some extent defined by the external fissures in higher primates.

The most primitive type of primate brain is exemplified by the treeshrew. The large olfactory bulb reflects its dependence on the sense of smell. The cerebrum lacks any fissures, and the specialized areas for sensation, voluntary movement and vision can only be defined microscopically.

In the more advanced monkey and ape brains the principal fissures delimit these specialized areas; the central fissure divides the area associated with voluntary movement from

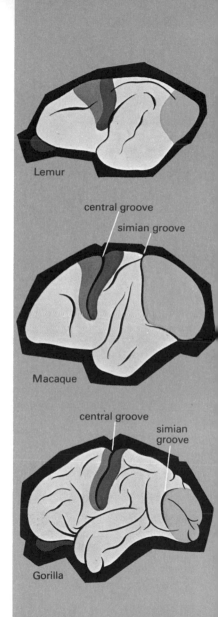

Diagram of brain of a treeshrew *(left)* indicating the positions of some major functional areas.

Brains of lemur *(top)* macaque *(centre)* Gorilla *(bottom)*, not drawn to scale, showing the changes in relative size and importance of some major functional areas.

Lemur

central groove

simian groove

Macaque

central groove

simian groove

Gorilla

the sensory area which receives impulses from skin, muscles and joints. The area receiving visual messages from the eye is defined by the conspicuous 'simian fissure', so-called because it is characteristic of monkeys and apes.

Lemurs have more advanced brains than treeshrews. The fissures of the cerebral cortex do not, however, outline the specialized areas. The specialized areas of the brain of the macaque are defined by conspicuous fissures. The brain of the Gorilla is further increased in size and in the complexity of the convolutions of the cerebral cortex.

Areas of the cerebrum between the main motor and sensory regions are known as 'association areas'; they integrate the activities of all parts of the cerebrum, acting as memory-storage units, and playing a co-ordinating role between perception of a stimulus and the voluntary response.

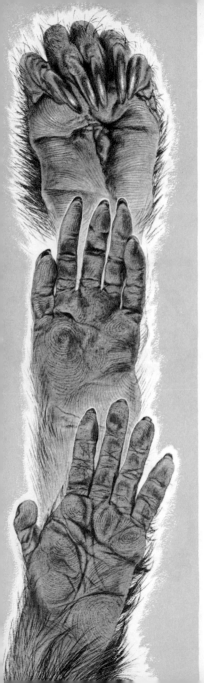

Hands

In arboreal life the force of gravity acts as a continued challenge to the tree-dweller. The prehensile or grasping hand of monkeys and apes has evolved primarily with the need for maintaining stability in trees. The essential features are long, strong, mobile fingers and opposable thumbs.

For lightly built primates such as marmosets, five digits with claws give an adequate hold when they are scurrying along large branches; the thumbs are not opposable. The hands of the New World monkeys have curved nails and the thumb is not fully opposable. The 'all-purpose' hands of macaques and other Old World monkeys can be used as feet when running or as a hand when grasping. By virtue of their fully opposable thumb, macaques can pick up small objects and perform finely co-ordinated movements such as grooming, or plucking blades of grass.

The huge hands of the

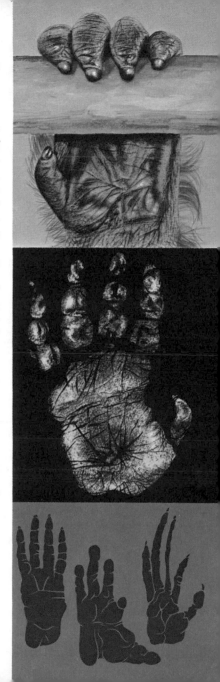

Orang-utan hand *(top)* in brachiating position
Hand print of Orang-utan *(centre)*
(bottom) Hand prints of spider monkey *(left)*, Potto *(centre)*, Aye-aye *(right)*

Orang-utans reflect their specialized arboreal way of life. The extremely long curved fingers are used to hook over branches while the animals hang and swing by their arms. Their thumbs are reduced in size and as a result manipulation is clumsy.

Spider monkeys' thumbs are absent or reduced to a tiny tubercle. The thumbs of Pottos are very large and set at an angle of 180° to the other digits, forming a pair of powerful pincers. The hand of the Aye-aye is unique; with its wire-like middle finger it can extract grubs from crevices in the bark of trees.

Apart from the treeshrews and marmosets, all primates bear true nails on their hands. Their function is to protect the soft sensitive pads at the ends of the digits. The skin of these pads is covered with fine ridges arranged in circular patterns (fingerprints). The digital pads are rich in sensory nerve endings which convey information to the brain about the shape, size and texture of objects.

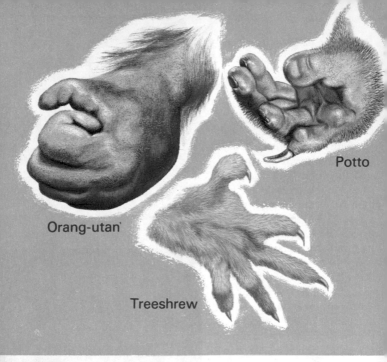

Orang-utan

Potto

Treeshrew

Feet of some arboreal primates

Feet

The hindlimbs of monkeys take a large share of weight-bearing when climbing and also provide the strong propulsive force necessary for making long leaps, while the forelimbs are subjected to less strain. The feet are therefore usually larger and more powerful than the hands; the big toes are widely separated from the other digits to give a firmer stronger grasp on the branches.

Even the small and primitive treeshrews show the beginning of the trend towards a divergent big toe. The foot of a Potto is bigger than the hand and bears a toilet claw on the small second digit. The big toe is widely separated from the other toes, and gives the foot a very strong grip. The macaque foot has a widely spread big toe which enables it to grasp branches firmly, thus giving the animal greater stability and agility when climbing. Apart from the treeshrews man is the only primate without an opposable big toe. The big toe of man has a different function, providing the propulsive force in his bipedal striding gait.

Orang-utan feet are functionally a second pair of enormous hands; they are about the same size, and perform the same strong grasping function. Like the thumb, the big toe is very small; in many individuals the big toe nail is absent.

In lemurs and lorises the toes bear flat nails but a claw (on the second digit) has been retained and is used as a scratching device or toilet claw which, together with the dental comb, is used in grooming activities. The tarsiers have retained a toilet claw on both their second and third digits and when scratching they bunch their feet into a fist from which the two toilet claws project. Marmosets, on the other hand, have a flattish nail on their rather short big toes, the rest of the digits being clawed. All other primates have true nails on all the digits though those of New World monkeys are rather curved and pointed.

Foot prints

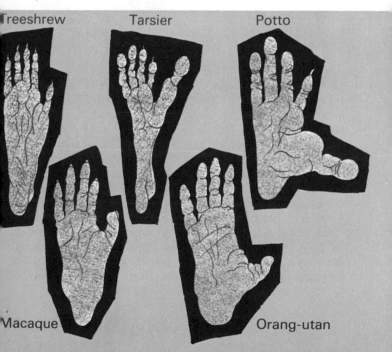

Treeshrew   Tarsier   Potto

Macaque   Orang-utan

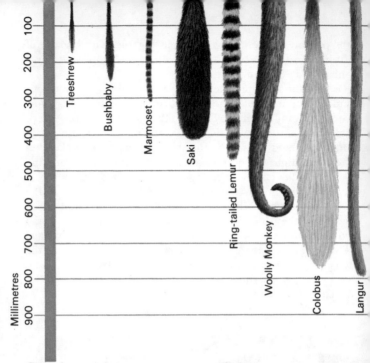

Tails of some primates

Tails

The majority of primate tails are adaptations to the acrobatic arboreal way of life. The tail can act as a balancing aid or counterpoise, a rudder, an air-brake, a support for the upright body and its most specialized function is to act as a grasping device – an extra hand. This is only found in a few of the New World monkeys. Generally tail length increases with the body length of the animal. On the other hand many primates have only an insignificant and useless stump in place of a tail.

By definition apes do not have tails; like man, they have only vestiges of tailbones, called the coccyx, at the base of the spine. Some monkeys, such as the Celebes Black Ape and the Barbary Ape, are so-called because they have no external tail. But they are not apes; in fact they are tailless monkeys. The only true apes, all of which are tailless, are the gibbon, the Orang-utan, the chimpanzee, and the Gorilla.

Prehensile tail of a woolly monkey, the bare skin on the underside is very sensitive and similar to the tips of fingers

Spider monkeys can feed hanging by their tails.

The spider monkey has one of the most remarkable tails of all primates – a grasping, or prehensile tail. The tail can support the whole weight of the animal while both hands are used for feeding. The tail-tip has an area of bare skin with sensitive 'fingerprint' ridges; its musculature is so finely co-ordinated that it can be used to pick up a peanut. This extra grasping ability is particularly useful as the spider monkey is thumbless and cannot pick up small objects very easily with its hands.

Other New World monkeys that have this type of prehensile tail are the howlers and the woolly monkeys. The capuchins have semi-prehensile tails that are capable of grasping but not of supporting their full weight. No Old World monkey has a prehensile tail, though some new-born animals have a brief period when the tail is semi-prehensile. This helps to provide the infant with a firm grip on its mother.

49

PRIMATE LOCOMOTION

The movements of primates, other than man, can be divided into three basic actions. They are grouped according to the type of movement most frequently used, though of course they can and often do move in other ways. These three groups represent the evolutionary trend in primate locomotion from the Eocene age to the present day, that is the lengthening of the arms in relation to the legs.

Tree-hopping – sometimes called vertical clinging and leaping – is found among prosimians, particularly the sifakas, Indris, Avahi, galagos and tarsiers; the legs are much longer than the arms. When the animal is clinging to a tree the legs are acutely bent and the body is held upright, supported by the arms. A leap is made by a sudden frog-like straightening of the legs; a turn through 180° can be made in mid-air. On landing the feet touch the tree-trunk first and thus take most of the strain.

Tree-hopping

Walking and Running

Arm-swinging

Quadrupedal running – this group includes animals that walk and run on all fours, either on branches or on the ground. As in fossil Pliocene monkeys the arms and legs are almost equal in length. Examples of this type of locomotion are seen in some prosimians such as the lemurs and in all New World and Old World monkeys. The slow, cautious climbing of the Potto and Slow Loris is a specialized form of quadrupedal movement.

Brachiation or arm-swinging is seen in the gibbons and the Siamang, more rarely in chimpanzees; a modified form is used by the Orang-utan. The arms are much longer than the legs and locomotion is by arm-swinging, the animal hangs by the arms below the branch. Chimpanzees and Gorillas are long-armed animals which usually walk on the ground; the knuckles of their hands support the body at an angle of 45° and the feet are flat on the ground. Knuckle-walking is probably an adaptation to ground living in animals too bulky for arm-swinging in the trees.

Successive stages in a leap of the tree-hopping locomotion of an Indris. The animal is mainly vertical. The line sketch illustrates the relative lengts of the arms and legs. All tree-hoppers have legs that are longer than the arms.

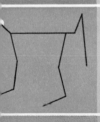

Stages in the gait of a baboon running along the ground. The body is horizontal and the arms and legs are about the same length.

Stages in the arm-swinging of a gibbon. In all the brachiators the arms are very much longer than the legs.

Skeleton and body outline of baboon

Upright posture

In the three-dimensional world of the forest, it is obvious that whilst climbing, the position of the body will be vertical as often as it is horizontal. Tree-living animals such as prosimians, monkeys and apes, have all developed a tendency towards the upright posture of the head and body.

In vertical clinging the body is held upright, while in arm-swinging the body hangs straight down. Even in quadrupedal runners the posture of the body when sitting is upright.

Animals which sleep sitting on narrow branches have special tough pads on their buttocks. These tough horny pads are attached to the ischial bones of the pelvis (the hip bones) and are known as the ischial callosities. They are insensitive and can support the full weight of a heavy animal, allowing it to sleep comfortably on narrow branches where it is safe from attacks by nocturnal predators. Ischial callosities are present

Baboon resting in tree in upright position

Ceylonese Grey Langur *(top)* can stand bipedally.
Woolly monkeys *(centre)* use their tails as a support, forming a tripod. Pinchés *(bottom)* may stand upright when excited.

in all Old World monkeys and in the gibbons and Siamang, but not, as a rule, in the great apes; they are occasionally present in chimpanzees.

Besides sitting upright many monkeys that usually move quadrupedally can stand and even run on two legs. When standing, the tail is often used as a support so that a tripod is formed by the legs and tail. The woolly monkey and the Patas Monkey may use the tail in this way. Pinchés, or Cottontops, also stand upright when alert or excited, rising from a sitting position.

Japanese Macaques can run and walk bipedally when carrying food in their hands. The upright stance also helps to increase the range of vision and is thus often seen in ground-living species; the Ceylonese Grey Langur stands briefly upright to peer over tall grasses, as does the chimpanzee in a similar situation. The Gorilla sometimes stands bipedally but mainly during chest beating and associated display activities.

53

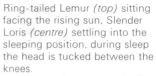

Ring-tailed Lemur *(top)* sitting facing the rising sun, Slender Loris *(centre)* settling into the sleeping position, during sleep the head is tucked between the knees.

Howler monkey *(bottom)* often the arms hang down on either side of the branch too.

Chimpanzee *(right)* sleeping in the characteristic position with the legs drawn up and lying on the crude nest of leafy branches

Resting postures

The primates without ischial callosities (the prosimians, New World monkeys and the great apes) have different positions in which they rest and sleep.

Ring-tailed Lemurs often sit upright and face the early morning sun, it is thought that they benefit from the ultra-violet light which is absorbed through the skin and lightly furred abdominal skin. Most lemurs sleep in a semi-erect position with the long bushy tail curved over their shoulders like a fur wrap.

Lorises and Pottos sleep with their bodies rolled into a ball. The Slender Loris, for example, sometimes chooses a forked branch in which to sleep, and then tucks its head

well down between its thighs and holds on to the branch
securely with both hands and feet.

Some monkeys sprawl along a branch, relaxing completely
with their arms and legs dangling. Howler monkeys, from
Central and South America, sleep in this position, the prehensile tail keeping a firm grip around a suitable branch.

The great apes often build crude nests of branches for
periods of sleep or daytime rest. Chimpanzees make a fresh
one every night, simply bending a few leafy branches together
to form a kind of platform, which is surprisingly strong. They
usually sleep lying on their sides with their knees drawn up
but they also lie on their backs. The nest takes up to five
minutes to construct and may be from 15 to 100 feet (5 to
30 metres) above the ground. Adult male Gorillas usually
sleep on the ground due to their great size and weight, but the
less bulky females and young may build their nests which
are rather crude platforms in the lower branches, up to about
20 feet (6 metres) from the ground. Nests may be made for
resting periods during the day too. Gorillas have no predators
other than man, and can therefore relax and sleep on the ground
safely. Orang-utans also build simple nests, but gibbons,
however, do not. They usually sleep in a furry ball, the head
tucked down between the knees and chest but often rest
sprawled on a branch in the daytime.

Hand function

In all primates the hand has three functions; firstly to support the body during locomotion; secondly to transfer objects such as food from one place to another; and thirdly to maintain contact with the environment through the special sense of touch. This description, however, could be applied to many non-primate mammals. The special signicance of the primate hand is the greater importance of the second and third functions.

The essential feature of primate locomotion is climbing which involves prehension or gripping; in this activity the branch is held between the folded-over fingers and the palm of the hand. In Old World monkeys the thumb is opposable and can thus be drawn away from the other digits to form – with the palm – the second jaw of a clamp. The posture of the hand of Old World monkeys when holding a branch is essentially the same as the posture in which man grips a hammer. In man this grip is called the power grip and has clearly been derived from the climbing grip of arboreal monkeys. Thus climbing, the essential ingredient of monkey locomotion, paved the way for the primate hand to develop as a prehensile organ capable of manipulating objects. The hands of New World monkeys and prosimians are not so well suited for manipulation, lacking a fully opposable thumb.

The evolution of an opposable thumb and sensitive pads at the tips of the digits made possible another type of grip, called the precision grip; here the object is held between the tips of the fingers and the thumb. It is particularly useful when small or fragile objects are manipulated, and when delicate tasks are being performed. Some Old World primates have better precision grips than others. Man is pre-eminent, the baboon is the next best and the apes are the least good.

The proportions of man's hand and his finely co-ordinated muscular system permit his grip to be both delicate and precise. Baboons have very short fingers and a long thumb so that their precision grip is quite well executed; both grass plucking and grooming involve the use of this type of grip. Chimpanzees have long fingers and a short thumb which interfere with the effectiveness of their precision grip. To hold fairly small objects the thumb opposes the bent fingers in a rather clumsy action.

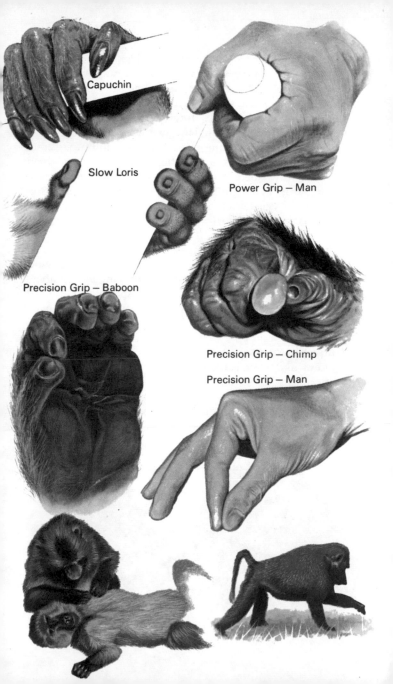

Capuchin

Slow Loris

Power Grip – Man

Precision Grip – Baboon

Precision Grip – Chimp

Precision Grip – Man

DIET

Most monkeys and apes are omnivorous, eating whatever can be found in the forests where they live. In tropical forests fruits, nuts and leaves are usually plentiful but are often supplemented with lizards, frogs, snails, insects, birds and birds' eggs. Primates living in grasslands have to be more resourceful; they dig in the ground for roots and tubers, overturn stones to find ants and beetles, and some more adventurous monkeys such as mangabeys, macaques and baboons raid orchards and fields of maize, rice and sugarcane. Some primates are vegetarian; but while eating fruits and leaves some insects are necessarily eaten too.

In tropical forests, where it rains almost every day, moisture is obtained by licking raindrops from leaves and bark and also from the juice of fruits. In the monsoon forests of India, where several months may elapse without rain, langurs are able to survive without drinking at all; their specialized digestive system enables them to get sufficient moisture from the mature leaves which form their basic diet.

Cheek pouches

Monkeys and apes cannot hoard or store food while it is plentiful, as squirrels and hamsters do, since their basic foods are perishable. However, like hamsters, some monkeys have developed large cheek pouches into which they can cram a

Baboon emptying food pouch *(left)*, Bonnet Macaque with distended food pouch *(right)*

	Insects Grubs Lizards etc.	Flowers Fruits Nuts Seeds	Leaves Stems Roots Grasses	Main Food Sources
Gorilla			■	Leaves
Orang-utan		■	■	Fruits
Chimpanzee	■	■	■	Fruits
Siamang		■	■	Fruits
Baboon	■	■	■	Fruits
Langur			■	Leaves
Howler			■	Leaves
Macaque	■	■	■	Fruits
Gibbon		■	■	Fruits
Spider Monkey		■		Fruits
Capuchin	■	■		Fruits
Lemur		■	■	Fruits
Potto	■	■	■	Insects
Galago	■			Insects
Squirrel Monkey	■	■		Fruits
Loris	■			Insects
Marmoset	■	■	■	Insects
Tarsier	■			Insects

Food sources of some primates

private store of food, which can be carried away and eaten in safety later. This is a useful adaptation for crop-raiding monkeys such as macaques. Cheek pouches are capacious extensions of the mucous membrane which lines the inside of the mouth; they may reach below the jaw but do not meet in the midline. After prolonged distension the muscular pouches become flaccid, and the food must be pushed back into the mouth using the hands.

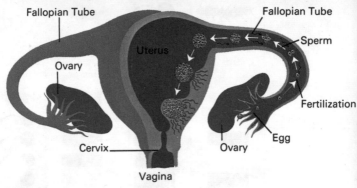

Female reproductive system

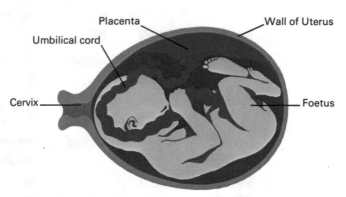

Human foetus just before birth

REPRODUCTION

In monkeys and apes reproduction is similar to that of the majority of mammals. During mating millions of sperms are deposited in the vagina of the female. Each sperm is mobile, able to swim with its powerful tail; some reach the fallopian tube where a single sperm fertilizes the egg. Then follows a long period of development – the gestation period.

The fertilized egg migrates to the uterus and becomes embedded in the wall where the placenta develops. In the placenta the bloodstreams of the mother and the embryo meet, separated only by thin membranes. Nutrients and oxygen pass through these membranes via the umbilical cord to the embryo and wastes pass out. At birth, the cervix slowly expands and the powerful muscles of the uterus contract expelling the infant still attached to the umbilical cord.

The female reproductive cycle

The ovaries contain many eggs, one or more of which ripen every month. The ripening egg is surrounded by the ovarian follicle which produces a hormone, oestrogen. After about fourteen days the follicle bursts, releasing the egg; this is termed ovulation. This first phase of the cycle – the follicular phase – is influenced by the concentration of oestrogen in the bloodstream.

The follicle, now referred to as the corpus luteum (yellow body) produces another hormone, progesterone which stimulates the internal lining of the uterus to develop a layer of tissue richly supplied with blood vessels in preparation for the fertilized egg. This second phase – the luteal phase – is influenced by progesterone in the bloodstream. If no fertilized egg reaches the uterus, the thickened lining is sloughed off and discharged as menstrual bleeding, approximately twenty-eight days from the beginning of the cycle.

Female Reproductive Cycle

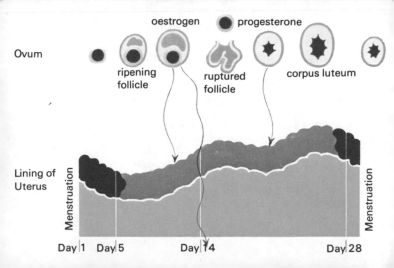

Ovum

oestrogen   progesterone

ripening follicle

ruptured follicle

corpus luteum

Lining of Uterus

Menstruation

Menstruation

Day 1   Day 5   Day 14   Day 28

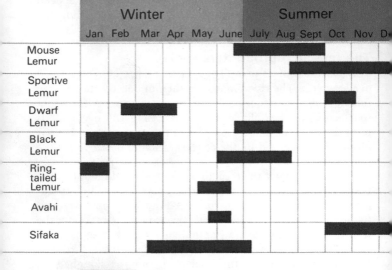

Mating Season (where known)

Birth Season

Seasonal breeding in some Madagascan primates

Oestrus

During the reproductive cycle, female monkeys and apes experience a period of oestrus which lasts about seven to fourteen days. At this time the female is particularly attractive to the male and actively seeks to mate with him. Oestrus occurs regularly, approximately every twenty-eight days, about the midpoint of the menstrual cycle, coinciding with ovulation.

Menstrual cycles of some monkeys and apes

New World monkeys		Old World monkeys		Apes	
Capuchin	18 days	Macaque	28-31 days	Gibbon	27-29 days
Woolly monkey	21 days	Guenon	30 days	Chimpanzee	35 days
		Baboon	35 days	Orang-utan	29 days
Spider monkey	24-27 days	Langur	30 days	Gorilla	30-31 days

Lemurs and lorises have a more primitive type of reproductive system; menstruation does not occur. All prosimians have oestrous periods but those that have been studied in the wild in Madagascar show a restricted mating season. No births have been recorded among primates in Madagascar in the first months of winter. Seasonal changes in climate and vegetation probably influence the onset of the mating season.

Sexual swelling

In some monkeys and apes changes appear in the skin surrounding the external genitalia of the female during oestrus. This area of skin – the sexual skin – deepens in colour and becomes distended. The swelling occurs during the follicular phase and is influenced by the oestrogen in the bloodstream. At ovulation the swelling usually reaches its maximum size; then in the luteal phase it gradually shrinks.

Only Old World primates show cyclical changes of the sexual skin, principally the baboons, mangabeys, Mandrills and some macaques. The chimpanzee is the only ape to show sexual swelling. The prominent and brightly coloured swelling probably acts as a visual signal to the male that the female is sexually receptive.

Female baboon with maximum sexual swelling, which coincides with ovulation. Mating thus takes place at the peak period of the cycle for fertilization.

Dry Season				Monsoon	
Jan	Feb	Mar	Apr	May	June
Gestation				Birth	Lactation

Breeding cycle of macaques in India

Mating

Almost all observations of menstruation in monkeys and apes have been on captive animals; the occurrence of menstruation in the wild is probably rare, as the female monkey is either pregnant or feeding her infant. For the female Rhesus Macaque in northern India the birth of an infant is an annual event; the year is wholly occupied with pregnancy and lactation.

During a period of oestrus, she seeks out one of the males of her group, indicating by the gesture of 'presenting' that she is ready to mate. They enter into a 'consort' relationship which may last from a few hours to several days; this relationship excludes other males, but sometimes includes the female's infant of the previous year, now about six or seven months old. The pair move and feed together, often grooming each other, and mating takes place frequently during this time. During oestrus the female may have a 'consort' relationship with several males in succession.

If fertilization occurs both menstruation and oestrus are suppressed during the gestation period of five and a half months.

Birth

Macaques appear to suffer the same stress and discomfort as is experienced by a woman in labour, but it is less prolonged. The infant's head (shown in blue in the diagram) fits the birth canal so tightly that the muscles of the uterus must contract very strongly to expel it. Normally the head is born first, followed a few minutes later by the rest of the body with the umbilical cord attached. Shortly afterwards the placenta is expelled, and is usually eaten by the mother who bites off the cord to within an inch or two of the infant's abdomen. Within

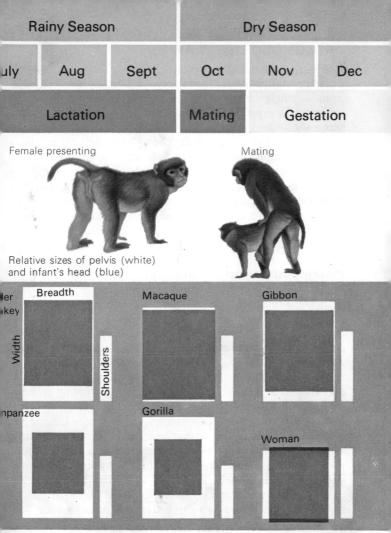

Rainy Season			Dry Season		
uly	Aug	Sept	Oct	Nov	Dec
Lactation			Mating	Gestation	

Female presenting

Mating

Relative sizes of pelvis (white)
and infant's head (blue)

Breadth

...ler
...key

Width

Shoulders

Macaque

Gibbon

...npanzee

Gorilla

Woman

an hour the infant's eyes are open and it is able to cling to its
mother. During its first day of life suckling is established.

The head of the great apes at birth is small in relation to the
female pelvis, so they have relatively easy births. The apparent
disproportion in humans is resolved by the infant's head
turning through 45° to enter the birth canal.

PROLONGED LIFE PERIODS

Prolonged gestation period

During gestation the fertilized egg – or zygote – must develop from a single-celled organism into a multi-celled animal capable of a separate existence. First the zygote undergoes a series of divisions into thirty-two cells. At this stage it becomes embedded in the uterine wall. The embryo now undergoes a series of complex changes, its cells increase in number and begin to assume different roles in the development of the organs, the muscles and the bones of the infant. The length of time that this process of differentiation, growth and development takes varies. More complicated animals need longer gestation periods and a progressive lengthening of gestation periods is seen in the primates.

Prolonged period of infancy

Infancy lasts from birth until the appearance of the first permanent tooth. Once again, the tendency is to prolong this period. During this time the infant is wholly dependent on its mother for its two principal needs – security and food. Interaction between mother and infant forms the basis of its social development.

The juvenile primate has a great deal to learn before it can play its part as an adult member of its group. Among ground-living primates, such as baboons, discipline is particularly necessary if the group is to survive in the somewhat rugged conditions of life in the open. In forest-living societies, such as chimpanzees, discipline is less strict because the environment is less demanding; nevertheless appropriate social behaviour is extremely important and must be learned by example and by trial and error. A protracted period of growth ensures long exposure to a particular code of behaviour.

Increased longevity

The increase in life span of higher primates ensures that the younger members of the group have elders able to pass on to them the experiences of a long life. The transmission of learning from the old and well-informed to the young and naive ensures the perpetuation of knowledge and experience.

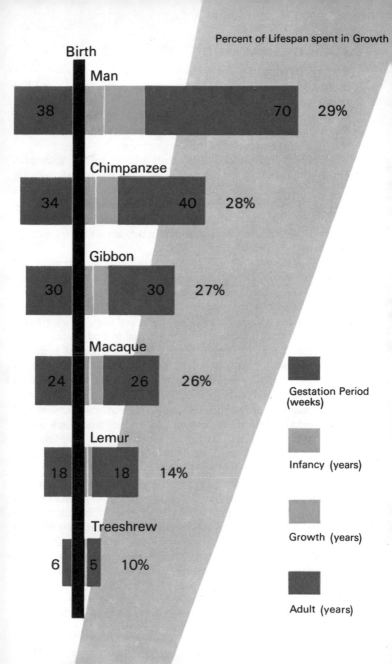

Percent of Lifespan spent in Growth

Birth

Man
38 | 70 | 29%

Chimpanzee
34 | 40 | 28%

Gibbon
30 | 30 | 27%

Macaque
24 | 26 | 26%

Lemur
18 | 18 | 14%

Treeshrew
6 | 5 | 10%

Gestation Period (weeks)

Infancy (years)

Growth (years)

Adult (years)

SOCIAL BEHAVIOUR

Social grouping

As in all mammals, the basic social unit is the female and her infant; adult males and females must also consort for the purpose of mating. Most primates form stable social groups. There are four basic types of grouping that can be related to the habits and habitats of various species.

All solitary primates are arboreal and nocturnal; examples are found among prosimians such as the dwarf lemurs, Sportive Lemur, Potto, lorises and the Aye-aye. Mouse lemurs and galagos sleep during the day in communal nests or 'dormitories'; their night-time activities in the wild are virtually unknown. Males and females are almost equal in body size.

The family group consists of an adult male and female and their offspring. This grouping is only seen among arboreal species, the sifakas and Indris from Madagascar, the titis, Douroucouli and marmosets from South America, and the gibbons from South East Asia. Males and females of species that form family groups are almost equal in body size.

The multi-male group is the commonest amongst primates and is found in both arboreal and ground-living forms. It consists of several adult males and about twice as many adult females, together with their offspring. This pattern is seen in some lemurs such as the Ring-tailed Lemur, many New and Old World monkeys and in Gorillas and chimpanzees. Males in multi-male groups are often much larger than females; this is particularly evident in ground-living forms such as the common baboons of East and South Africa.

The one-male or harem group is an adaptation of the multi-male group to arid environments. Several large males in a group would eat a disproportionate amount of the scanty food supply. As only one male is necessary to father the next generation, the one-male group is the most efficient reproductive unit in this environment. The harem principle is seen in the ground-living Patas Monkey, Hamadryas Baboons and Gelada. Gelada one-male groups often congregate in herds of as many as 400 animals at feeding sites and sleeping cliffs. Adult males in such groups are twice the size of females.

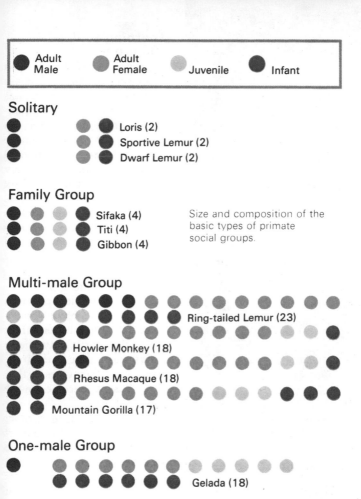

Adult Male   **Adult Female**   **Juvenile**   **Infant**

Solitary
Loris (2)
Sportive Lemur (2)
Dwarf Lemur (2)

Family Group
Sifaka (4)
Titi (4)
Gibbon (4)

Size and composition of the basic types of primate social groups.

Multi-male Group
Ring-tailed Lemur (23)
Howler Monkey (18)
Rhesus Macaque (18)
Mountain Gorilla (17)

One-male Group
Gelada (18)

Field studies of apes and monkeys have only been undertaken within the last forty years, and are still in progress. The information shown in the chart is based on a limited number of observations. Many species are completely unknown, with regard to their social behaviour. A detailed study of a species in the wild requires several years of careful observation.

69

Mother baboon and infant

Learning to live in the group

A primate group is basically a reproductive unit, but in ground-living species such as baboons it may also be a community that finds safety in numbers and in defensive display. A baboon group consists of approximately forty animals of which about six are large adult males. These larger animals form a nucleus of protection, a kind of bodyguard, for the females and young.

Maternal influence

A baboon mother provides food and a mobile 'home-base' from which her infant can venture forth, only briefly at first, to explore its environment; later it ventures further away, as it becomes more independent.

The mother is in contact with her infant for almost twenty-four hours a day; she feeds it, grooms it, plays with it and to some extent controls its activities – by pulling it back by the tail, for instance, when it strays too far away.

The infant soons learns to respond to her calls and

gestures; particularly to run quickly to her when danger threatens and the group must move rapidly to safety.

For the first four or five months of its life, it rides underneath its mother, clinging on to her fur with hands and feet, and often feeding from the nipple at the same time. Later, it rides on her back like a jockey, leaning against her tail and clings to her back fur with its hands.

Influence of adults
Quite soon after birth, a baboon infant and its mother become the centre of attraction for the entire group. The biggest adult males walk beside the mother as she carries her infant, or sit beside her when she is feeding it. Adult females and juveniles groom the mother and attempt to groom the infant. Thus it grows up surrounded by attentive relatives and becomes almost inevitably, a completely socialized being. The adult males, usually aloof, forbidding and strict disciplinarians, are completely tolerant of the young infant as it plays near them or scrambles over them.

Influence of age-group
As the infant begins to move away from its mother for longer periods, it becomes strongly attached towards other infants of the same age-group. Most births occur at a certain peak-period during the year, usually just before the rainy season,

Adult male and female baboons with young

Baboons playing

when food is most abundant. Thus all infants are about the same age and gradually spend more and more time playing together. All sorts of climbing, chasing and wrestling games help to develop the bodily skills needed for the future.

Appropriate social attitudes and gestures are learned at this stage; the 'presenting' of the hindquarters as a sign of appeasement or submission, and the mounting of another animal to indicate dominance are two of the most important as they fore-shadow the mating behaviour of adult life.

By two years of age male and female play-groups have separated; young males play rough boisterous games on the edge of the group; young females spend more time in grooming and in quieter play in the centre among the adult females and their babies. During the rough-and-tumble of play, an infant may be hurt and cry out in pain; the adult males intervene instantly to protect the young animal.

Facial darking in young baboons

Growing-up

At birth the baboon infant has a black coat of fur, with pink hands and feet, a pink muzzle and large pink ears. It is in this initial period, from birth to three months, that the greatest interest is shown in the young animal by other members of the group.

From four to six months the skin begins to darken; the coat colour gradually changes to fawn. Adult females, other than the mother, pay less attention to the infant, but it still receives protection from the adult males.

When the infant is between seven months and a year old adult coloration is gradually achieved. Weaning takes place between twelve and fifteen months, and thereafter the young juvenile is almost completely independent. Females become sexually mature at three and a half years and have their first baby at about four years. Males become sexually mature at four years, but are not fully grown until about eight years old.

Many other primates have a distinctive colour during early infancy; this draws the attention of the group to the newborn infant and its mother who need the maximum protection at this time.

Adult male baboon disciplines a troublesome juvenile by neck biting

Adult male hierarchy

A baboon group may consist of about six adult males, twelve adult females and twenty-four juveniles and infants of both sexes. Adult males sometimes develop a hierarchy in which the strongest, toughest animal is the undisputed leader; the next strongest ranks immediately below and so on down a graded hierarchy to the weakest. More often, however, social factors, rather than sheer strength, influence the selection of the leader.

The discipline exerted by the dominant males is important for the integrity of the group; it ensures that the females are mated by the dominant males when at the peak of their oestrous cycle – the time when they are most likely to conceive – thus maintaining the genetic 'fitness' of the group.

External dangers, such as attacks from predators, are met by the strongest and most experienced animals.

Within the group, peaceful relations are maintained; fights and squabbles are controlled and the weaker animals protected. The discipline exerted by the leader, though sometimes aggressively enforced, lessens conflict within the group.

Grooming

Social grooming ia a unique primate activity and it plays an important part in the social life of baboons and other primates. During the resting periods of the day, one baboon will approach another and present itself (or a particular part of its body) for grooming. The other animal responds by parting and examining the fur, picking off dry skin, dirt and parasites, either by nibbling or with the finger and thumb. This is repeated many times, the fur being parted systematically over a small area so that the whole is carefully scrutinized and cleaned. Judging by the relaxed attitude of the groomee, it is a pleasurable process; and the groomer, by her close attention and concentration, expresses interest and satisfaction in the task. Grooming is often reciprocal; the roles are interchanged and groomer becomes groomee. Thus, friendly social relations are established and maintained between animals in the group.

Several hours of every day are spent in grooming. All adult animals groom but females groom the most; they groom infants, juveniles, adult males and each other. Grooming also has a valuable hygienic function in keeping the skin and fur clean and free from parasites; incidental cuts and wounds receive particular attention.

Social grooming

Group defence

A baboon group comprehends a pool of knowledge and experience which exceeds that of any individual member and enables it to deal successfully with the hazards of life on the ground.

In their daily search for food, baboons must move long distances in open grassland. Here they are exposed to sudden attack by predators such as Lions, Leopards, Cheetahs, hyenas, jackals and wild dogs.

When moving to a new feeding area baboons adopt a defensive grouping shown in the upper diagram in simplified form: the dominant males stay in the centre of the group to protect the females and infants. Subordinate and young males lead and bring up the rear, the positions of greatest danger.

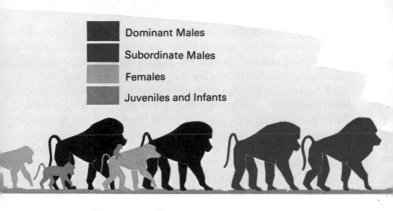

Dominant Males
Subordinate Males
Females
Juveniles and Infants

If, for example, a Leopard is encounted in the long grass the leading males give a loud bark of alarm. This warning signal alerts the whole group; the dominant and subordinate males move forward towards the predator, while the females and their young retreat rapidly. This is indicated in the lower diagram. The fierce aspect of several large male baboons with their heavy manes bristling and their long canine teeth exposed may be enough to discourage a predator from attacking, at least until the females and young have reached a safe distance. The relatively weak females and young are thus protected by the large and stronger males.

In such a dangerous situation, communal behaviour has obvious survival value; a straggler from the group would be an easy prey.

Defensive positions

Home range

The term 'home range' generally describes a living area which is not defended by its occupants. Baboons live within a 'home range' in which each bush, tree and rock is a familiar landmark and which includes certain key-points needed for survival; food sources, a river or waterhole, and trees or cliffs for sleeping. These frequently used sites are referred to as 'core areas'.

Depending on the abundance of food and the size of the group, the home range may vary from 2 to 12 square miles. The daily movements of the group within this area may be from a few yards up to 12 miles (19 kilometres); the daily average is about 3 miles (5 kilometres).

Because of the scarcity of water sources in some grassland areas, many baboon groups must often share a waterhole; around it, therefore, there is a considerable overlap of home ranges. Here baboon groups converge peaceably, usually ignoring or avoiding each other. An example of home range overlap is seen in the diagram below.

Home ranges of Olive Baboon groups in Nairobi National Park, Kenya

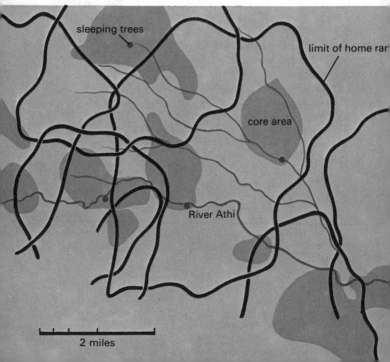

Baboons and other mammals at a waterhole

Baboons are also seen in the company of other animals such as zebra, gazelles and wildebeeste. Their attitude is one of mutual tolerance, but it can also become one of mutual dependence. In a mixed group of baboons and Impala, for instance, the baboons' keen eyesight supplements the Impala's sense of smell and acute hearing to prevent surprise attacks by predators. Impala may also depend on baboons to protect them from attacks by Cheetahs.

Sleeping sites
A safe refuge for sleeping at night is of the utmost importance to a ground-living species. Well before darkness falls, baboons move towards the trees or cliffs where they will spend the night. Before the sun sinks they are safely ensconced among the slender branches where predators cannot follow them without giving plenty of warning of their approach.

Treeless habitat

Territory

In tropical rain forests, fruit, leaves, insects and water are relatively plentiful. Here certain species live in well-defined 'territories' which are defended; they will not allow others of the same species to enter their established domain.

The original definition of territoriality applied to song birds but it applies equally well to gibbons which exhibit, to some degree, all four aspects of the original definition. Advertisement: all gibbon families start the day with very loud high-pitched wailing calls which advertise their whereabouts to the neighbouring gibbon families. Intolerance: these signals stimulate neighbouring families to approach and confront each other on the boundary, or 'no man's land', between their two territories. Here a ritualized conflict occurs, the two males hooting, swinging about in the branches and chasing each other. Very occasionally one animal may actually be bitten, but usually the conflict ceases after about thirty minutes without injury on either side, and the families separate. Isolation: these conflicts keep gibbon families isolated. Fixation: gibbons are not nomadic and remain confined within their territorial boundaries which are, however, not absolutely rigid and show some flexibility.

The gibbon family consists of an adult male and female, and up to four offspring. Gibbons are monogamous and have an infant on average every two years. The father shows some interest in the infant, protecting it, playing with it and grooming it. The young gibbon becomes sexually mature at about seven or eight years of age; by this time there may be two or three younger gibbons in the family. The young adult is encouraged to leave the group by the increasing antagonism of the parents. Adult gibbons feel hostility for other adults of the same sex, whether they come from neighbouring groups or from within the family. The young adult is ejected from the group and lives alone until it finds a mate who has been similarly expelled. The chances of finding a mate are increased by the close proximity of territories.

Other primate species showing territorial defence are the Sifaka, Ring-tailed Lemur and the Titi. Howler monkeys do not defend any fixed boundary but rather the place where they happen to be at the time.

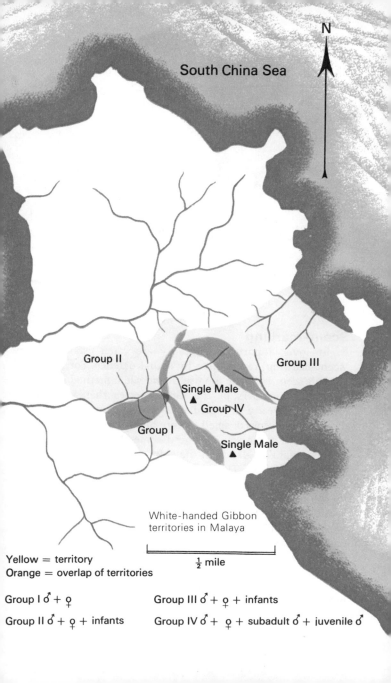

N

South China Sea

Group II

Group III

▲ Single Male
Group IV

Group I

▲ Single Male

White-handed Gibbon
territories in Malaya

½ mile

Yellow = territory
Orange = overlap of territories

Group I ♂ + ♀

Group II ♂ + ♀ + infants

Group III ♂ + ♀ + infants

Group IV ♂ + ♀ + subadult ♂ + juvenile ♂

Golden Lion Tamarin scent-marking *(left)*
Slender Loris urine-marking *(right)*

Scent-marking

Rights over a particular locality are often established by 'scent-marking' rather than by actual defence. For nocturnal animals scent-marking is a useful and silent method of informing other members of the same species that a particular locality is already occupied. It is observed mainly in prosimians whose sense of smell is more highly developed than that of monkeys and apes. They and some New World monkeys have specialized glands in the skin which produce a strong-smelling secretion. These glandular areas are rubbed against branches, thus spreading the scent. For example the Golden Lion Tamarin marks branches with secretions from a scent gland in the neck as do the treeshrews, Indris and sifaka.

Urine may also be used for marking. Galagos, of both sexes, wet their hands and feet with urine, then spread the scent as they move about among the branches. Slender and Slow Lorises also mark branches with urine. They may either excrete a few drops at a time as they slowly creep along a branch, or they may wash their hands and feet in urine.

Although they are diurnal the Ring-tailed Lemurs have

an intricate system of scent-marking, which plays an important part in the complex social life of the species. They have a raised horny scent gland on the forearm which is more highly developed in males than in females, and a smaller gland is found near the armpit of males. These glands are used in what have been called 'stink-fights' between males. These fights are associated with the establishment of male dominance within the group.

At the start of one of these fights the contestants rub their forearm and chest glands together. The long, ringed tail is then impregnated with the scent by being passed over the horny gland on the forearm. Arching the tail forwards over his head the male wafts the smell towards his opponent. The dominant animal gradually advances, forcing the subordinate's retreat.

Both male and female Ring-tailed Lemurs also mark branches with a secretion from the genital region.

Ring-tailed Lemur in stink fight. Transferring scent from arm and chest glands on to tail *(top)*, wafting scent towards opponent *(centre)*, female marking with clitoral gland secretion *(bottom)*

Very young Slow Loris 'parked' on a branch.

Infant care

Among the prosimians a few species give birth to immature young in nests. These include the treeshrews, and some of the lemur family. Other prosimians are sufficiently mature at birth to be able to cling to the mother's fur. Some infants are left 'parked' on a branch while the mother searches for food. All infant monkeys (but not all apes) are capable of grasping the mother's fur and supporting themselves almost from the moment of birth.

Gorilla and chimpanzee infants are relatively weak and unable to cling unaided. They are supported by their mothers for the first two or three months. Then they ride on the mother's back. Like the Slow Loris, a chimpanzee mother may also hang her infant up on a nearby branch for a short time while she is feeding or nest building.

Langur running with hour-old-infant clinging to her belly.

Male marmoset carrying twins

Paternal care

Marmosets are thought to live in family groups (an adult male and female and their offspring). Twins are usual and are carried by the father which only hands them over to the mother for breast feeding. This behaviour is also seen in the titis and the Douroucouli.

In family groups the adult male is inevitably the father of the offspring. In multi-male groups, however, the actual father cannot be known for certain as a female may mate with all the males in her group. Nevertheless, male Japanese Macaques have been observed to show a form of 'paternal' care towards an infant which is by chance deprived of its mother's care. Although there may be no true paternal link, the adult male looks after the infant in exactly the same way as its mother would.

Adult male Japanese Macaque grooming a year-old-infant

A 'motherless' Rhesus Macaque mistreats her new-born infant, crushing it on the floor.

Maternal behaviour

From the foregoing outline of social behaviour, it can be seen how great an influence the mother has on the development of her infant. She is, with a few exceptions, wholly responsible for its upbringing and early training.

As part of a study of learning and intellectual development, infant Rhesus Monkeys were separated from their mothers and hand-reared in isolation. Naturally it was hoped that they would eventually breed.

These 'deprived' monkeys were completely anti-social, sitting passively in their cages, or compulsively rocking themselves back and forth. Out of nearly 150 monkeys studied, not one 'deprived' male succeeded in mating. Only after perseverance on the part of non-deprived experienced males were four of the females finally mated.

Moreover these four 'deprived' females showed abnormal maternal responses. They ignored their infants, brushing them away when they tried to feed, striking and biting them.

Further studies have shown that appropriate behaviour in adult monkeys can only be acquired through a combination of maternal care and the company of other young monkeys during the early months of infancy.

Infant affection

The tie that binds the infant monkey to its mother is its first link with its social environment; the relative importance of various aspects of that tie have been studied extensively.

Macaque infants were separated from their mothers at birth and placed in cages with a variety of dummy mothers. One was made of wire mesh with a nipple on the front from which milk could be sucked; another was made of soft terry-towelling, but had no milk supply. Infants tested with these two devices were found to spend as much time as possible clinging to the soft warm mother, leaving it reluctantly to feed from the wire mother. It appears that bodily contact – something soft and warm to cling to – is the principal element in the macaque infant's needs in the early stages of its life.

The soft mother also appeared to supply a sense of security to the infant. When confronted with a frightening situation – an animated toy for instance – the infant rushed to its terry-towelling mother, appearing to derive comfort and confidence from its warmth and softness.

Infant clings to the soft substitute mother but feeds from the wire mother

Dominant Rhesus Macaque

Communication

All primates need to communicate, and a primary need is that species should recognize each other. Faces (and sometimes the buttocks and genital region) of many monkeys show strongly marked colour flashes and/or stripes. Vivid eye-patches, ear-tufts or moustaches help the animal instantly to recognize its own kind.

Prosimians rely less on visual signals and more on olfactory signals to transmit this vital information. Primates, living in the tropical forest or savanna where the field of view is limited, supplement their visual channels of communication with vocalization in order to keep in touch with other members of their group.

Compared with prosimians, monkeys and apes have a wider range of facial expression and vocalization which, together with body posture and gestures, should be considered as part

Low ranking Macaque

Macaque threat gestures *(right)*

of a complex signalling system which conveys information about mood or intention to other group members.

Many multi-male groups are controlled by a single dominant male. For example the Rhesus Macaque controls his group not so much by aggression as by the threat of it. He can threaten an offender by a series of signals of graded intensity. The threat may be averted at any time by a submissive gesture on the part of the offender. Thus the threat of aggression, which is conveyed by well-understood signals, may actually prevent fighting. The gestures of the dominant animal are reinforced by sharp coughing grunts. If no submissive response is received, the final gesture of lunging at the offender is followed by chasing, catching, striking and biting him.

The dominant male demonstrates his superiority to all the other males in the group by his calm bearing and sleek appearance, his tail is carried high. The lowest ranking male, on the other hand, is lean and scruffy; he walks with his legs slightly bent and carries his tail low.

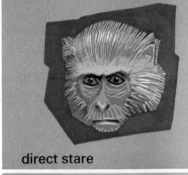

direct stare

open mouth

head bobbing

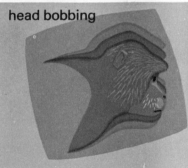

ground slapping

Gorilla chest-beating

Macaque tree-shaking

Baboon yawn

Displays

Various types of display behaviour have been described, some concerned with the defence of territory and others with preserving the social dominance order.

Some of the threat gestures and facial expressions described for macaques can be seen on other species such as baboons. Elaborate yawning for example, which reveals the large canine teeth, is a threat display usually directed at other individuals within the group but it may be used against predators. Slapping the ground with the hand is also seen in baboons.

When two groups of Japanese Macaques approach each other, adult males climb to the tops of trees and shake them violently. Branch-shaking is seen in other macaques and branch-breaking occurs during group encounters in colobus monkeys and gibbons.

Howlers react to the appearance of other howler groups or any intruder such as an observer with loud gurgling roars and the breaking off and dropping of dead branches. Faeces and urine, too, are often released when the howler is immediately above the observer which gives the impression that the dropping of excrement is intentional rather than random.

Chimpanzee groups react excitedly when they meet; they whoop and shriek noisily; they shake and break off branches; they wave and throw them; they thump the ground and drum on the buttress-roots of trees with their feet.

In the Gorilla, a display of this kind has become 'ritualized'; that is to say the Gorilla responds to a particular situation in a stereotyped way. The motivation of the Gorilla's chest-beating display is not clearly understood but it is thought to be an expression of inner tension, in the presence of another group, when the Gorilla is undecided whether to attack or retreat. A series of hooting calls is followed by the placing of leaves or a branch in the mouth ('symbolic feeding'); the Gorilla rises, plucking and throwing handfuls of vegetation. The ritual culminates in standing erect, beating the chest with the cupped hands, crashing through the undergrowth, uprooting saplings, branch-shaking and thumping the ground with the hands. Beating the chest is performed by Gorillas of all ages except the very young; however, only the dominant adult male performs the full ritual.

RECOGNITION SECTION

The principal characters of the more common species, broad geographical range and brief notes on ecology and habits are given in the following section. For a brief classification see pages 8 to 11. The measurements given are those for an adult male of the species, of average size.

Treeshrews (Tupaiidae)

Although the general appearance of treeshrews is not typical of primates, they share many primate characters particularly in the structure of the skull and the brain. The most squirrel-like treeshrews *(Tupaia, Anathana* and *Urogale)* range throughout South East Asia. The Madras Treeshrew *(Anathana ellioti)* is confined to southern India; the Philippine Treeshrew *(Urogale everetti)* is found on Mindanao Island in the Philippines. Between these two extremes, treeshrews of the genus *Tupaia* range from the Himalayas to the Palawan Islands east of Borneo, from high up in the mountains of China to sea-level in the Malaysian archipelago.

These treeshrews have a longish muzzle with naked moist snout, naked ears, claws on all five digits of hands and feet and a bushy tail. The coat colour is speckled due to the light and dark banding of individual hairs, and varies from buff to red-brown with an oblique pale stripe on the shoulder.

The smooth-tailed treeshrews *(Dendrogale)* are more mouse-like owing to their small size and their smoothly furred tail. They are found in southern Indochina and in Borneo. They do not have a pale shoulder stripe.

All treeshrews have thirty-eight teeth (see page 41). They eat insects, grubs and beetles as well as some fruit and plants and all are forest-living. Only the Feather-tailed Treeshrew *(Ptilocercus lowii)* is nocturnal. This habit is indicated by its larger eyes, shorter muzzle and larger, more mobile ears. The naked scaly tail with a tuft distinguishes it from the other treeshrews. It is found in Malaya, Sumatra, Borneo and some of the adjacent islands.

Top to bottom: Feather-tailed Treeshrew, Common Treeshrew, Smooth-tailed Treeshrew, Philippine Treeshrew

Ruffed Lemur

Ring-tailed Lemur

Lemurs (Lemuridae)

The lemurs are confined to Madagascar. The five species of true lemurs *(Lemur)* live in the dry deciduous and tropical rain forests. They are diurnal and appear to be wholly vegetarian. Lemurs have rather dog-like faces with a longish muzzle, a naked moist nose and large well-furred ears. But the long bushy tail, the mobile limbs, the grasping hands and feet and the digits bearing nails proclaim that they are primates. The second toe bears a 'toilet' claw used in conjunction with the dental comb for grooming purposes.

True lemurs

The Ruffed Lemur *(Lemur variegatus)* is slightly larger than the others. Its young, often twins or triplets, are reared in a nest and suckled mainly at night. Male Black Lemurs *(L. macaco)* are entirely black, but females are reddish-brown with white cheeks and ear tufts. The Red-bellied Lemur *(L. rubriventer)* is red-brown with red underparts, and a black tail. The Mongoose Lemur *(L. mongoz)* is grey-brown; the males have red cheeks and females white cheeks and throat.

Grey Gentle Lemur,

Mongoose Lemur-female

male

Gentle lemurs

The gentle lemurs *(Hapalemur)* live in lakeside reedbeds as well as in the forests. They are diurnal and feed on plant foods including the pith of reeds. Their gait is of the tree-hopping type; they have been observed grasping an upright reed in a vertical clinging position. The male Grey Gentle Lemur *(H. griseus)* has a two-inch strip of glandular skin on the fore-arm. The slightly larger Broad-nosed Gentle Lemur *(H.simus)* is now thought to be extinct.

Sportive lemur

The Sportive Lemur *(Lepilemur mustelinus)* differs from all the other lemurs in being nocturnal and in having thirty-two teeth, all the other lemurs having thirty-six; it lacks upper incisors. It forms a small cluster of about twenty solitary animals. The largest group consists of two members, the female and her offspring, with the male a few trees away. Sportive Lemurs have large eyes and large oval ears, indicating their nocturnal habit. The coat is grey with a reddish tinge; there may be a dark stripe down the centre of the back.

Dwarf and mouse lemurs

The dwarf lemurs and mouse lemurs are arboreal and nocturnal denizens of the tropical rain forests of Madagascar. They have a short muzzle with a naked moist nose, large eyes with dark rings round them, and large mobile ears. A scratching claw is present on the second toe; all the other digits have nails. They have thirty-six teeth with a characteristic dental comb formed from the lower incisors and canines.

The Fat-tailed Dwarf Lemur *(Cheirogaleus medius)* is so-called because of its ability to store fat at the base of the tail, prior to periods of hibernation which take place during the winter (May to August). It becomes torpid and sleeps for several weeks. The Greater Dwarf Lemur *(C. major),* shows a similar tendency but its periods of torpor last only two to three days. *Cheirogaleus* are mainly fruit eaters but the Fork-marked Dwarf Lemur *(Phaner furcifer)* is more insectivorous. The dark mark down the middle of its back divides on the head, to meet the dark eye rings.

Mouse lemurs *(Microcebus)*

Greater Dwarf Lemur *(above)*
Fat-tailed Dwarf Lemur *(below)*

are very attractive-looking furry animals, but they are shy, wild and virtually untameable. They, too, show some seasonal lethargy but this is less marked than in the dwarf lemurs. There are two species: the Lesser Mouse Lemur *(M. murinus)* has a white streak down its short pointed muzzle, while the larger Coquerel's Mouse Lemur *(M. coquereli)* lacks the white nose streak.

Microcebus murinus is one of the smallest primates: other contenders for the title of smallest primate are the Dwarf Galago *(Galago demidovii)*: $5\frac{1}{2}$ inches (14 centimetres) long with tail $7\frac{1}{2}$ inches (19 centimetres); the Pygmy Marmoset *(Cebuella pygmaea)*: $5\frac{1}{2}$ inches (14 centimetres) long with tail 8 inches (20 centimetres); the Pygmy Treeshrew *(Tupaia minor)*: 5 inches (12 centimetres) long with tail 6 inches (15 centimetres); the Smooth-tailed Treeshrew *(Dendrogale murina)*: 5 inches (12 centimetres) long with tail 5 inches (12 centimetres); and the Tarsier *(Tarsius)*: 5 inches (12 centimetres) long with tail 8 inches (20 centimetres).

Fork-marked Dwarf Lemur *(above),* Mouse Lemur *(below)*

Verreaux's Sifaka *(left)* Indris *(right)*

Indrises (Indriidae)

Indrises are also found only on Madagascar. They are closely related to the lemurs but have only thirty teeth. They have only two premolars on each side instead of the usual three (see p. 41) and the lower canines are absent. The dental comb therefore consists of only the four lower incisors. Another striking characteristic is the length of the legs in relation to the arms, which is related to their tree-hopping gait (see p. 50). The very large foot with its huge opposite big toe is used in grasping vertical tree trunks. A scratching claw is present on the second toe.

The Indris *(Indri indri)* is restricted to the tropical rain forest areas. It lives in family groups and has an eerie wailing song. Its name is derived from a misunderstanding on the part of one of its first European observers. A native called out to him 'Indris! Indris!' which he assumed to be the native name of the animal. In fact it meant 'Look at that!'

The Indris and sifakas *(Propithecus)* are both diurnal and

rather similar except that the Indris is larger and has a mere stump of a tail whereas the sifakas have long tails which, though not prehensile, are often curled up like a watch spring. The sifakas are mainly white or pale grey but the face is black and there is a dark patch on the crown. Sifakas live in family groups and are vegetarian.

The Avahi *(Avahi laniger),* sometimes known as the Woolly Lemur on account of its dense fluffy coat, is much smaller and is nocturnal. Coat colour is a neutral grey-brown with a lighter area around the eyes which gives it an owlish look.

Aye-aye (Daubentonidae)

The Aye-aye *(Daubentonia madagascariensis)* is one of the most extraordinary primates. It is a solitary nocturnal animal living on grubs and fruit such as lichis and coconut in the tropical rain forests of Madagascar. Its teeth (only eighteen in number) are remarkable for the absence of canines, the reduction in the number of molars and the chisel-like incisors. The Aye-aye has large eyes, large mobile ears, a pointed muzzle, a long coarse blackish coat and a long bushy tail. All the digits have claws except for the big toe which bears a flat nail.

Aye-ayes use their chisel-like incisiors to gnaw bark and expose the grubs; the wire-like middle finger is used to remove grubs from their holes.

Both Slender Lorises *(left)*, and Slow Lorises *(right)* are slow moving. They grip the branches with their pincer-like hands and feet

Lorises (Lorisidae)

The lorises differ from the lemurs in certain significant characters of the skull, but they resemble them in having a naked moist nose, thirty-six teeth with a similar dental comb and a scratching claw on the second toe. All members of the loris family are arboreal and nocturnal but they can be divided into two distinct behavioural types, the slow-moving, solitary lorises and quick-moving galagos or bushbabies.

Watching a loris or Potto move is like seeing a film in slow motion. It appears to flow along the branch; it will even proceed spirally around the branch to avoid projections. This stealthy prowling is thought to aid them in capturing insects, lizards and nestlings which form a large part of their diet.

Lorises and Pottos

The Slender Loris *(Loris tardigradus)*, from southern India and Ceylon, has a short pointed muzzle, large eyes and ears, extremely thin wrists and ankles and no tail. A white streak down the nose separates the dark pear-shaped eye patches. The Slow Loris *(Nycticebus coucang)*, from other areas of South East Asia, is very similar but its ears are smaller, its limbs more robust, and the dark line down the back sometimes extends to the ears and to the dark patches around the eyes.

The Potto *(Perodicticus potto)*, from East, Central and West Africa, lacks any dark markings and has a short tail. Behind the neck, hidden by fur are three or four bony spines, covered with skin. The Angwantibo *(Arctocebus calabarensis)*, slightly smaller, has neither the bony spines nor tail.

Galagos

Galagos form the second group in the loris family and are found over a wide area of tropical Africa, in the rain forest and woodland savanna. Their social habits have been little studied but they appear to sleep in communal nests or 'dormitories' and may be quite gregarious. Like the lorises, they eat a large proportion of insects and animal food.

They are superficially rather similar to the dwarf lemurs of Madagascar but have relatively longer legs. Their locomotion is of the tree-hopping type. The Bushbaby *(Galago senegalensis)*, which is only about 6 inches (15 centimetres) long, with a tail of 9 inches (22 centimetres), can jump over 7 feet (2 metres) vertically from a standing start. Up in the trees, they make huge leaps from branch to branch; moving over the ground from one clump of trees to another, they proceed in bounding hops, using both legs together, like a kangaroo.

The smallest of the galagos, the Dwarf Galago *(G. demidovii)*,

Potto *(left)*, Dwarf Galago *(right)*

Fat-tailed Galagos *(left)* are found in most of Africa south of the Sahara. Bushbabies *(right)* live in equatorial Africa.

is really very like a mouse lemur with a similar short pointed muzzle; however the legs, particularly in the ankle region, are much more elongated.

The biggest galago, the Fat-tailed Galago *(G. crassicaudatus),* is about the size of a rabbit. It has a long dog-like muzzle and very large mobile ears. Galagos can move their ears independently back and forth to catch the faint sounds of insects which form a major part of their diet.

Other galagos are similar in appearance but are of different sizes; Allen's Galago *(G. alleni),* from equatorial West Africa, has a head and body length of about 8 inches (20 centimetres), with tail 10 inches (25 centimetres), and the Bushbaby is slightly smaller. The Needle-nailed Galago *(G. elegantulus)* is so-called because its nails have a raised ridge running down them which ends in a sharp point. The head and body measure 9 inches (22 centimetres), the tail 12 inches (32 centimetres).

Galagos have broad expanded soft pads on the tips of the fingers which are sharply bent at the middle joint. This helps to give them a good grip on the branches.

Tarsiers (Tarsiidae)

The tarsiers have sometimes been called 'living fossils'. The proportions of the limbs, which indicate their tree-hopping gait, are very similar to those of early primates of the Eocene period. Their legs are very much longer than their arms, and the ankle, or 'tarsal' region, is particularly elongated, giving rise to their name. There are three species from Sumatra and Borneo, Celebes and the Philippines.

Tarsiers differ from the galagos in having a dry furry nose instead of a moist grooved one. The teeth (thirty-four in number) also differ in that the lower incisors are reduced to one pair, and there is no dental comb. However, tarsiers have a scratching claw on their third as well as their second toes (see page 47). Enormous eyes and large ears indicate their nocturnal habit. The base of the long tail is sometimes used as a support; it is naked except for a fuzz of fine hairs at the tip.

The tarsiers can turn their heads through 180° and look directly backwards, and can turn the body through 180° in mid-air when leaping. These adaptations help them in catching their food which consists mainly of insects, lizards and spiders.

Tarsiers have discs on their finger tips, to increase their grip. Their triangular nails are very small.

Marmosets (Callitrichidae)

The marmosets are a family of tiny New World primates which includes the true marmosets, the tamarins and Goeldi's Marmoset. They differ from all other primates in having modified claws on their hands and feet; these claws, though long and pointed, are slightly open on the underside, so they are rather more nail-like than claw-like. The big toe is small and bears a flat nail.

Marmosets have thirty-two teeth, the same number of teeth as the Old World monkeys, the apes and man. But, whereas the Old World monkeys (and apes and man) have two premolars and three molars, marmosets have three premolars and two molars. The cebids have three premolars and three molars (see page 41). Marmosets' noses are dry, the nostrils are well separated and face sideways. The tail is not prehensile.

True marmosets

The true marmosets *(Callithrix)* are distinguished from the tamarins by their elongated lower incisor teeth, as

Common Marmoset *(top)*.
Black-tailed Marmoset *(bottom)*

tall as the canines. They are found in the tropical rain forest and dry deciduous forest of Brazil. They are arboreal and diurnal, eating insects as well as some plant foods. They are thought to live in monogamous family groups; the female usually has twins which the male carries on his back.

The Common Marmoset *(C. jacchus)* has ear tufts which surround and sometimes completely conceal the bare ears; the tufts may be black or white. The coat is mottled greyish-brown due to the light and dark banding of individual hairs. The Santarem Marmoset *(C. humeralifer)* has tufts growing from the edge of the ears. The Black-tailed Marmoset *(C. argentata)* has bare pink ears and a bare face. There are dark and pale races of the two latter species; one race of each is almost completely white.

The Pygmy Marmoset *(Cebuella pygmaea)* is found in Brazil, Ecuador and Peru. Its black and buff banded hairs produce an overall brownish speckled effect with vague ringing of the tail.

Santarem Marmoset *(top)*,
Pygmy Marmoset *(bottom)*

Saddle-back Tamarin *(left)*, Red-handed Tamarin *(centre)*, Black and Red Tamarin *(right)*

Tamarins

All the eleven species of tamarins *(Saguinus)* live in the tropical rain forests of Central and South America. They are arboreal and diurnal, feed on insects, fruit and seeds, and are thought to live in family groups.

Their general appearance is similar to the marmosets but they are a little larger and the tail is never banded. Their lower canine teeth are much longer than their incisors.

The coat colour is extremely variable, with many striking colour patterns, marking and adornments. Tamarins can be divided into two main groups: those with hairy faces and those with bare faces.

The hairy-faced tamarins include a species in which both the facial hair and skin are black, as are the large ears; their arms and shoulders are also black while the back and rump are mottled buff and black. The Red-handed Tamarin *(S. midas*

midas) has red hands and feet, but the Negro Tamarin *(S. midas tamarin),* has black hands and feet.

In other hairy-faced species the facial skin is black but with short white hairs growing around the mouth. The animal has the appearance of holding a ball of snow-white cotton wool in its teeth. The Black and Red Tamarin *(S. nigricollis)* has two colour zones on its body, the black foreparts gradually shading into the dark red of the rump and legs. The Saddle-back Tamarin *(S. fuscicollis)* has three colour zones. The foreparts and hind parts contrast sharply with the mottled back. All the races have white hairs around the mouth but the body fur ranges from black and dark brown to reddish, orange or yellow. One race is almost white and some have white eyebrows like the one shown here.

The remaining hairy-faced species have white facial skin beneath the white moustache hairs. Three species are illustrated here and show a graded series in which the amount of white around the nose and mouth increases, culminating in the long drooping moustaches of the Emperor Tamarin *(S. imperator).*

White-lipped Tamarin *(above),*

Moustached Tamarin *(below)*

The Emperor Tamarin is named after Emperor Franz Joseph of Austria, he had magnificent moustaches.

The second group of tamarins – the bare-faced tamarins – have nevertheless a good deal of very fine hair on the face. Only the Pied Tamarin *(S. bicolor)* is truly bare-faced with black bare skin extending right up to the top of the crown between the large bare ears. This species is found in northern Brazil.

Geoffroy's Tamarin *(S. oedipus geoffroyi)* and the Pinché *(S. oedipus oedipus),* from Panama and north-western Colombia respectively, belong to a species which has tiny white hairs all over a black face, forming fine white lines across the brows and cheeks. The Pinché has a crest of long white hairs on the crown, giving rise to its common name of 'Cottontop'.

Although the White-footed Tamarin *(S. leucopus)* from north-western Colombia has the face and forehead covered with a considerable amount of silvery hairs, it looks bare-faced, especially in profile, because the crown hairs are much longer and thicker than the facial hairs.

The Golden Lion Tamarin *(Leontopithecus rosalia)* is instantly recognizable by its brilliant golden colour and its silky fur; the crown hairs sweep back from the brow, concealing the ears. One race is completely golden; the other two are black and gold.

Left to right Goeldi's Marmoset, Pied Tamarin, Geoffroy's Tamarin,

It can be distinguished from the other tamarins by certain skull characters, and by its very long narrow hands and feet; the fingers are extremely elongated and partially webbed. This rare tamarin lives up in the mountains of south-eastern Brazil. Because of its beauty it has been a target for hunters, but lately its export has been banned.

Goeldi's Marmoset

Although considered to be closely related to the tamarins, Goeldi's Marmoset *(Callimico goeldii)* contradicts one of the main characteristics of the family. It has thirty-six teeth, like the Cebidae, with three premolars and three molars on each side of the jaw. However the teeth are distinctly tamarin-like and, considering its other characters, such as the clawed hands and feet, it is placed in the marmoset family.

This rare little animal has only been found on some south-western tributaries of the River Amazon. Goeldi's Marmoset is entirely black except for a minute white tip on the end of each hair. The crown hairs are swept back off the face and the bridge of the nose is depressed.

Golden Lion Tamarin, White-footed Tamarin

The New World monkeys (Cebidae)

The New World monkeys of Central and South America include the squirrel monkeys, Douroucoulis, titis, sakis and uakaris, with non-prehensile tails; and the capuchins, spider monkeys, woolly monkeys and howlers with prehensile tails. They are arboreal, and all are diurnal except the Douroucouli.

The sakis and uakaris live in the tropical rain forests; and are thought to feed mainly on fruit. Their noses are the most extreme examples of the platyrrhine condition, the nostrils being very widely separated. The nails, though long, curved and pointed, are true nails. Like other cebids they have thirty-six teeth but with unusually protruding incisors and projecting canines.

Sakis and bearded sakis

The sakis *(Pithecia)* are distinguishable from the bearded sakis *(Chiropotes)* by the broad bulging bridge of the nose; the thick tail tapers to a point. White-faced sakis *(P. pithecia)* show some sex-related differences. The infants resemble one parent as soon as they are born. Both sexes of the Monk Saki

White-faced Saki family, female carrying young

Bearded Saki *(left)*, Red Uakari *(centre)*, Monk Saki *(right)*

(P. monachus) resemble the female White-faced Saki.

The bearded sakis are found both north and south of the River Amazon in Brazil and in the Guianas. The bridge of the nose is narrow but the nostrils are very widely separated. The tail is also long and thick but does not taper to a point. The Black Bearded Saki *(C. satanas)* has a brown back with black head, limbs and tail while the White-nosed Bearded Saki *(C. albinasus)* has a white patch on the nose and lips.

Uakaris

The Bald Uakari *(Cacajao calvus)* is similar to the Red Uakari *(C. rubicundus)* but has silvery-white fur. The Black-headed Uakari *(C. melanocephalus)* is brown with black extremities; the crown and forehead in this species are furred.

Widow Monkey *(left)*,
Douroucouli *(centre)*,
Dusky Titi *(right)*

Douroucouli

The Douroucouli or Night Monkey *(Aotus trivirgatus)* is the only nocturnal monkey. It is widely distributed in Central and South America. It lives in family groups, nesting during the day in hollow trees. The eyes are very large, the nose small and pointed; the nostrils are closer together than is usual in New World Monkeys. It is easily identified by the three dark lines on the forehead, separated by a white patch over each eye.

Titis

The diurnal titis *(Callicebus)* are rather similar but have much smaller eyes and the nose is typically broad and flat. They live in family groups and show some territorial defence. Like the Douroucouli and the marmosets, the male titis often carry the offspring. Mated pairs twine their tails together when sitting. The Dusky Titi *(C. moloch)* has banded hairs giving a speckled effect to the back. The Widow Monkey *(C. torquatus)* has a white throat patch and white hands that are reminiscent of the white scarf and gloves worn by widows in Latin American countries – hence its common name. A third species, from south-eastern Brazil, the Masked Titi *(C. personatus)*, is a speckled grey-brown colour, with black mask, black hands and feet and a red-brown tail.

Squirrel monkeys

The squirrel monkeys *(Saimiri)* are found in the tropical rain forests of Central and South America. They are easily recognized by their white faces which contrast with the dark crown, eyes and dark area around the mouth. In the Common Squirrel Monkey *(S. sciureus)* the crown of the head is approximately the same colour as the back, i.e. a speckled yellowish-grey, but in the Red-backed Squirrel Monkeys *(S. oerstedii)* the crown is black and the back bright orange-red. In both species the tail is greyish with a black tip.

Squirrel monkeys live in large multi-male groups of up to 500 animals. They feed mainly on insects and fruit. A single infant is born which looks exactly like its parents from the moment of birth; it is cared for entirely by its mother.

Common Squirrel Monkeys *(above)* and Red-backed Squirrel Monkey *(below)* are highly gregarious; a group of 500 has been seen.

White-throated Capuchin *(left)*,
Black-capped Capuchins *(right)*
have a dark band in front of the
ears.

Capuchins

The capuchins are found throughout Central and South
America. They have the typical platyrrhine nose. The tail is
not strictly prehensile but is used to curl round branches for
support. It is fully furred throughout. Capuchins live in small
multi-male groups, feeding on fruits and insects.

Capuchins of the same species and from the same locality do
not always look exactly the same. This makes identification
difficult, but broadly speaking they can be divided into
a 'tufted' and an 'untufted' group. The 'tufted' group includes
the Black-capped Capuchin *(Cebus apella)*. It has long dark hairs
on the crown which form tufts or ridges, or sometimes a mat
of erect hairs.

The 'untufted' group have no tufts on the crown and there
is no dark band in front of the ears. The White-throated
Capuchin *(C. capucinus)* has a cap of black hairs which con-
trasts with the white face, throat and shoulders. The Brown

Pale-fronted Capuchin *(left)*, Weeper Capuchin *(right)*

114

Pale-fronted Capuchin *(C. albifrons)* is similar but its colours are brown and cream. The Weeper Capuchin *(C. nigrivittatus)* is medium brown with darker extremities.

Howlers

Howlers are found throughout Central and most of South America. They live in tropical rain forest and montane forest, feeding mainly on leaves. They live in multi-male groups and show some territorial defence, roaring loudly at an encroaching group. They always have a rather hunched appearance. The prominent larynx, more developed in the male, bulges beneath the chin. The nostrils are set rather close together. The fully prehensile tail is constantly in action as they move rather deliberately through the trees.

The Red Howler *(Alouatta seniculus)* is a brilliant copper-red; the Brown Howler *(A. fusca)* a dark dull reddish-brown, and the Red-handed Howler *(A. belzebul)* black with red hands, feet and tail-tip. The Mantled Howler *(A. villosa)* is black, sometimes with a golden fringe along the flanks. In the Black Howler *(A. caraya)* the females and young are olive-buff, the males becoming black on reaching maturity.

Red Howler Monkey

Black-handed Spider Monkey *(left)*,
Long-haired Spider Monkey *(right)*

Spider monkeys

Spider monkeys *(Ateles)* live in tropical rain forest and
montane forest. They are found in Mexico further north than
the howlers; but their southerly limit is in central Bolivia.
They live in multi-male groups, and feed mainly on fruit.

Their most distinctive character is the reduction of the
thumb which is either completely absent, or reduced to a
small stump. The tail is prehensile and is in constant use as an
extra hand as they move rapidly through the trees. The arms
are longer than the legs.

Coat colour and texture are very variable, but the fur is
usually long and the crown hairs are directed forwards over
the forehead. The Black Spider Monkey *(A. paniscus)* may
have a striking pink face. The Brown-headed Spider Monkey
(A. fusciceps) is black with a brown head, or, may be completely
black. The body of the Black-handed Spider Monkey *(A.
geoffroyi)* may be either gold, red, buff or dark brown, while
the crown, hands, feet, knees and tail-tip are black. The
Long-haired Spider Monkey *(A. belzebuth)* is black or brown,
with paler underparts.

Woolly monkeys

The woolly monkeys *(Lagothrix)* look like rather chubby spider monkeys but possess a thumb. They have short, dense, plushy fur with long shaggy fur on the rather protruding abdomen; in South America, woolly monkeys are sometimes known as barrigudos or 'pot bellies'. Woollies may be grey, brown or black; usually the head and extremities are darker than the back. A rare species, Hendee's Woolly Monkey *(L. flavicauda)*, found in a remote Andean valley, is a deep mahogany with yellow on the underside of the tip of the prehensile tail.

The Woolly Spider-monkey *(Brachyteles arachnoides)* has the robust build and dense fur of the woolly, but the thumb is absent as in spider monkeys. The arms are longer than the legs, and the tail is prehensile. This large, very rare animal is confined to the tropical forests of south-eastern Brazil where the expansion of agriculture threatens it with extinction. An adult male can be 25 inches (62 centimetres) long with a tail 30 inches (75 centimetres) long. Coat colour ranges from grey to yellowish-brown; the nostrils are set rather close together.

The rare Woolly Spider-monkey *(left)* is only found in the forests of south-east Brazil. The Woolly Monkey *(right)* occurs in most of the Amazon basin

Old World monkeys (Cercopithecidae)

The second main group of anthropoids, the Old World monkeys can be divided into two subfamilies: firstly the omnivorous guenons, mangabeys, baboons and macaques (the Cercopithecinae) and secondly the leaf-eating langurs and colobus monkeys (the Colobinae). They are all diurnal and are widely distributed throughout Africa and Asia. They are found in a great variety of habitats including tropical rain forest, temperate forest, open grasslands and sub-desert.

Two main external characters distinguish Old World monkeys from New World monkeys; first, the nose is generally narrow with the nostrils facing outwards and downwards; and second, ischial callosities are present on the buttocks. They have flattish nails on all digits. Thirty-two teeth are present, the same number and kinds as in apes and man; the infants have twenty milk teeth, as man does. The tail is never prehensile.

The omnivorous Cercopi-

Vervet Monkey *(top)*, L'hoest's Monkey *(centre)*, Preuss's Monkey *(bottom)*

thecinae eat all kinds of foods and many raid crops and orchards. All have a simple stomach and possess cheek pouches where they can store small quantities of food for short periods.

Guenons

The guenons *(Cercopithecus)* are a very widespread genus of the omnivorous group; they are found throughout Africa south of the Sahara Desert. The ischial callosities are small and well separated.

The Savanna Monkey *(C. aethiops)* is a mainly ground-living species, living in woodland, savanna and sub-desert habitats. This species is called the Green Monkey in West Africa, the Grivet in Ethiopia and the Sudan, and the Vervet in South Africa.

Preuss's Monkey *(C. preussi)* and L'Hoest's Monkey *(C. lhoesti)* are two closely related species from eastern Nigeria and eastern Congo respectively. White-throated Guenons *(C. albogularis)* are found in most of eastern Africa; two races are illustrated. De Brazza's Monkey *(C. neglectus)* from the Congo

White-throated Guenons from South Africa *(top)* and Mount Kenya *(centre),* De Brazza's Monkey *(bottom)*

basin has very distinctive colouring. The Diana Monkeys from West Africa, with their chestnut backs, white thigh stripe, their snowy chests and black faces are among the most beautiful of all African mammals. There are two races, the Roloway Monkey *(C. diana roloway)* has a long white beard, while the beard of the Diana Monkey *(C. diana diana)* is short.

The Golden-bellied Guenon *(C. pogonias)* from Cameroun and Fernando Poo has three characteristic black stripes on the crown. The closely related Mona Monkey *(C. mona)* has a similar distribution on the mainland but extends westwards to Ghana.

Hamlyn's Owl-faced Monkey *(C. hamlyni)* has a distinctive pale stripe running from the forehead down the nose, and a diffuse pale band on the forehead. The body fur is speckled olive grey with black underparts. It is found in the eastern Congo.

There are two types of white-nosed guenons; those with an oval nose-spot and those with a heart-shaped nose-spot. The Greater White-

Top to bottom: Greater White-nosed Guenon, Lesser White-nosed Guenon, Redtail, Red-eared Guenon

nosed Guenons *(C. nictitans)* have a large white or whitish oval spot on the nose; the body fur is black speckled with grey or red-brown. They are found in West Africa from Liberia to the Congo River. The Lesser White-nosed Guenon *(C. petaurista)* has a white heart-shaped nose-spot and a diadem of pale speckled hairs on the crown; it is found from Gambia to Ghana.

The Redtail *(C. ascanius)*, is found south of the Congo River, from Angola to Uganda; it has a conspicuous white heart-shaped nose-spot, bushy cheek tufts and whiskers. The body fur is dark brown and the tail red-brown.

The Red-eared Guenon *(C. erythrotis)* has a red triangular nose-spot and red tufts on the ears. The body fur is black speckled with yellow or grey and the tail is bright red. Red-eared Guenons are found on Fernando Poo, and in Nigeria and Cameroun.

The newborn infants of most of the guenons, as well as those of baboons, are a distinctly different colour from their parents.

Top to bottom: Golden-bellied Guenon, Roloway Monkey, Hamlyn's Owl-faced Monkey, Mona Monkey

Allen's Swamp Monkey *(left)*, Moustached Monkey *(right)*

Another strikingly marked guenon is the Moustached Monkey
(C. cephus) which is found between the Congo and Sanaga
Rivers in West Africa. The violet-blue face has a white bar
on the upper lip below which is a fringe of black hairs giving
a 'moustache' effect. The body fur is speckled greenish-
brown and the tail reddish.

Allen's Swamp Monkey *(C. nigroviridis)* is a rather rare
animal; it differs from the other guenons in being slightly
smaller but it is more robust and has a relatively short tail.
The ischial callosities are large and sometimes meet in the
midline, in the males. Sexual swelling is seen in the females at
oestrus, an unusual character only seen in the Swamp Monkey
and the Talapoins among the genus *Cercopithecus*. Allen's
Swamp Monkey comes from the swamp forest areas of the
Congo basin; nothing is known of its life in the wild.

Another guenon, the tiny Talapoin Monkey *(C. talapoin)*
comes from the mangrove swamps and swamp forests of
equatorial Africa. They are found in large groups of eighty
to a hundred along river banks. They raid the native root-crop
of manioc from riverside pools where it is placed by the
natives to soak and purify. Predominantly greenish-grey,

Talapoins have pale underparts and a characteristic black smudge on the cheeks. Like Allen's Monkey, the females exhibit sexual swelling at oestrus.

Patas Monkey

The Patas Monkey *(Erythrocebus patas)* is found in Africa from the edge of the Sahara to about 3° south of the Equator. It lives in an open habitat which may be woodland savanna, grassland or sub-desert. It is ground-living as its long arms and legs of equal length attest.

The Patas Monkeys can run extremely fast and may cover long distances every day in the search for food. They live in one-male groups; the male acts as a watch-dog, advancing ahead of the group, occasionally climbing trees to spy out the land. Patas Monkeys adopt the habit of rising up on their hind legs to increase their range of vision. When alarmed, the Patas freezes in the deep grass to escape observation.

The male is considerably larger than and twice as heavy as the female. She does not exhibit sexual swelling at oestrus. The aquiline nose and prominent white moustache gives it a marked resemblance to an army officer of the 'old colonel' type, so it is also called the Military Monkey or the Red Hussar.

Talapoin *(left)*, Patas Monkey *(right)*

Mangabeys

The mangabeys *(Cercocebus)*, are found in tropical forests of Africa. Some species are partially ground-living and are notorious crop-raiders. Mangabeys are slightly larger than guenons, with long limbs and tails; they have hollow cheeks and a more prominent muzzle. Cheek pouches are present. The large ischial callosities are united in the males only. Sexual swelling occurs in females at oestrus.

The White-collared Mangabey *(C. torquatus)* comes from between the Niger and Congo Rivers. The Sooty Mangabey *(C. atys)* from further west, has a flesh-coloured face with dark patches. The Black Mangabey *(C. aterrimus)*, from south of the River Congo, has crown hairs forming a crest. The Grey-cheeked Mangabey *(C. albigena)*, from Central Africa is mainly black with long brownish hairs on the neck and shoulders. The grey cheek hairs are very short. The Agile Mangabey *(C. galeritus)*, of Central Africa and Kenya, is a light mushroom colour without any marked facial adornments.

Drill and Mandrill

Drills and Mandrills form a very distinctive genus from equatorial West Africa. They are very large, mainly ground-living monkeys with a short stumpy tail. Their most striking feature is the massive muzzle with big bony ridges on either side of the nose. In Mandrills *(Mandrillus sphinx)* the muzzle is bright blue; in the Drill *(M. leucophaeus)* the whole face is black. In both species the coat is sombre in colour, dark brown, charcoal grey or olive green. The females are considerably smaller than the males. Sexual swelling occurs at oestrus. Ischial callosities are united in the males but separate in the females.

Grey-cheeked Mangabey *(left)*, Black Mangabey *(centre)*, Sooty Mangabey *(right)*

White-collared Mangabey

Drill *(left)*, the facial mask of the Mandrill *(right)* matches the genital skin of females.

Baboons

Baboons are mainly ground-living monkeys and are found in woodland savanna, grassland and sub-deserts of Africa south of the Sahara and in south-west Arabia. The Olive Baboon *(Papio anubis)* ranges well into the Sahara Desert.

Baboons resemble Mandrills in being very large; the males are much larger than the females. The prominent muzzle with its bony ridges is black, except in the Hamadryas Baboon *(P. hamadryas)* which has a flesh-coloured muzzle. Individual hairs of the coat are coarsely banded with yellow and black forming an effective camouflage. The arms and legs are of almost equal length. In most species the tail is carried upright but at about one third of its length it makes a sharp downward kink. This tail posture is most useful as a support when the infant is carried on the mother's back. Sexual swelling is conspicuous in the females at oestrus.

Baboons are omnivorous, even eating small mammals such as hares or young gazelles. In South Africa they are said to

Olive Baboon

Female Hamadryas Baboon
(left), large male *(right)*

prey on young lambs when food is scarce. Meat-eating however is unusual and the staple diet is fruit, leaves, grasses and
roots.

Baboon species can be divided into two socially differing
groups. The typical baboons live in multi-male societies while
the Hamadryas Baboons live in one-male or harem societies.
The Olive Baboon is found from the Sahara and the Ivory
Coast to Kenya. The Yellow Baboon *(P. cynocephalus)* from
further south, is longer legged, with conspicuous white
cheeks. In southern Africa the darker, more heavily-built
Chacma Baboon *(P. ursinus)* replaces it. The Guinea Baboon
(P. papio), a reddish species, is found in West Africa.

The Hamadryas Baboon, from Ethiopia and south-western
Arabia, is socially and structurally distinct. The Hamadryas
group consists of a single adult male with one to four females,
together with their offspring. The adult male keeps his harem
together by strong disciplinary measures which include
biting his females on the nape of the neck. Many one-male
groups gather at night at the sleeping cliffs.

Gelada

Geladas *(Theropithecus gelada)* are easily distinguished from baboons. The large mane, the tufted tail, the equal length of the limbs and the relatively small size of the females are all Hamadryas-baboon-like characters. But the shape of the face is entirely different; the nostrils are small and the cheeks are hollowed, giving the face a 'figure-of-eight' appearance.

Both males and females have a patch of bare skin on the chest; in the females this is surrounded by small raised white blisters. The sexual skin of the female mimics her chest patch. Both areas of skin become bright scarlet during oestrus. Two tough-skinned hairless pads lie below the ischial callosities which are separate in both sexes.

Geladas live along the edges of rocky gorges in the mountains of Ethiopia. They gather at night in herds of 300 or 400 animals when seeking the safety of their steep sleeping cliffs. They forage in one-male groups when food is scarce and stay in large herds if food is plentiful.

Macaques

Macaques are the most widespread genus of monkeys. They live in all kinds of habitats in Asia, the Far East, North Africa and Gibraltar.

Male *(left)* and female *(right)* Geladas

Crab-eating Macaque

They are similar to the guenons but the muzzle is slightly longer and the brows more prominent; the cheek pouches are very large, a useful adaptation for these accomplished crop-raiders. In all macaques, except the Moor Macaque *(Macaca maurus)*, the ischial callosities are well separated.

Macaques can be grouped into long-tailed, medium-tailed and short-tailed species; there are even two tailless species.

Long-tailed macaques come from the most southerly part of their range. Crab-eating Macaques *(M. fascicularis)* are found in most of South East Asia as far east as Timor. The Bonnet Macaque *(M. radiata)* and the Toque Macaque *(M. sinica)* from southern India and Ceylon respectively, can be distinguished by the cap of long hairs on the crown. Crab-eating

Toque Macaque *(left)* and Bonnet Macaques *(right)* showing crown hair patterns

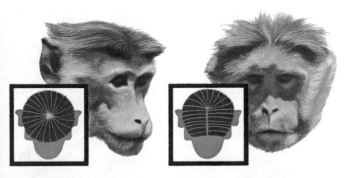

Macaques have a small sexual swelling at oestrus, while Bonnet and Toque Macaques have none at all.

Among the macaques with tails of medium length is the very beautiful Lion-tailed Macaque *(M. silenus)* which lives in southern India. This shy, forest-living animal has a slightly tufted tail and a striking grey ruff encircling the face. The sexual swelling of the female is not very marked. The Pig-tailed Macaque *(M. nemestrina)* is found in South East Asia, Sumatra and Borneo. The tail is carried arched over the back. The crown hairs of Pig-tails radiate from a central whorl on top of the head, giving it its characteristic horse-shoe shaped crown patch. The females have a very large and conspicuous purplish-pink sexual swelling at oestrus.

Other species with medium-length tails are the Rhesus Macaque *(M. mulatta)* with a 10 inch (25 centimetres) long tail from Afghanistan, northern India, Burma, China, northern Thailand and Vietnam; the Assamese Macaque *(M. assamensis)* has a tail 11 inches (27 centimetres) long; and the Formosan Rock Macaque *(M. cyclopis)*. At oestrus, the Rhesus and Assamese Macaques show reddening of the sexual skin, while the Formosan Macaque shows an extreme form of swelling that involves not only the root of the tail but the thighs as well.

The two short-tailed macaques are very different from each other although superficially rather similar, both having short

Japanese Macaque
Opposite: Lion-tailed
Macaque *(left)*, Pig-
tailed Macaque *(right)*

tails, long shaggy hair and red faces. The 4-inch-long (10 centimetres) tail of the Japanese Macaque *(M. fuscata)* is clearly visible as it is well haired with a tuft at the tip, while the 3-inch (7 centimetres) tail of the Stump-tailed Macaque *(M. arctoides)* is completely bare, bent to one side, and concealed in the rump fur. Japanese Macaques have thick yellowish-grey coats while the Stump-tail is rather thinly clad in long dark brown or grizzled hairs, particularly sparse on the under-parts. The forehead and cheeks are well covered with fur in Japanese Macaques, while the Stump-tail's face is crimson and the forehead becomes bald in adults. The pink faces and sexual skin of the females become red at oestrus in both species.

Stump-tailed Macaque
and infant

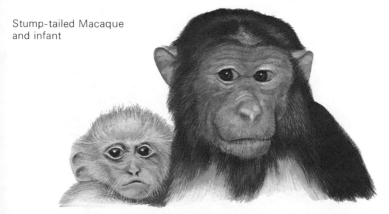

The two remaining macaque species are virtually tailless, though sometimes a small tubercle may be present. It is interesting that these two species are found at the extreme ranges of the distribution of the genus, in North Africa and in Celebes.

The North African species, called the Barbary 'Ape' *(M. sylvanus)* on account of its tailless condition, is a husky, hardy animal from the mountains and wilder parts of Morocco and Algeria. The macaques of Gibraltar, the only free-living monkeys in Europe, cannot be considered a wild population as they were introduced by man. During the Second World War a number of animals were imported to keep the small population up to strength. The fur, coarsely banded in black and yellow, gives a camouflage effect to the coat as it does in baboons. A conspicuous bluish swelling of the sexual skin indicates that the female is in the oestrous phase of her sexual cycle.

The Moor Macaque *(M. maurus)* from the southern peninsula of Celebes and nearby islands has a mere tubercle for a tail about 2 inches (5 centimetres) long. These animals are black or dark brown, sometimes with paler limbs and buttocks. The brow ridges are stronger than in other macaques, and the Moor Macaque male is exceptional in having the ischial callosities united rather than separate. At oestrus the females exhibit a prominent pink sexual swelling.

Barbary Ape *(left)*, Moor Macaque *(right)*

Celebes Black Apes, male *(left)*
female *(right)*

Celebes Black Ape

The Celebes Black Ape *(Cynopithecus niger),* from the northern peninsula of Celebes and some small offshore islands, is very closely related to the Moor Macaque but shows some striking differences.

It has a prominent crest on the crown in both sexes. The muzzle is broad and flat, due to the raised bony ridges on either side of the nose. Facial expressions include raising the eyebrows and retracting the scalp which simultaneously depresses the crest. Baring of the teeth and gums causes the skin over the bony nose ridges to become wrinkled. The brow ridges are very prominent. The ischial callosities consist of two pairs of smooth hard pink pads, closely surrounded by the dark fur. Females exhibit a prominent pink sexual swelling at oestrus. The fur is either black or dark brown.

This concludes the Cercopithecinae, the subfamily of omnivorous Old World monkeys. Their ability to eat all kinds of food and their general adaptability has enabled them to survive and breed in zoos and they are, therefore, better known than the other subfamily, the Colobinae, the leaf-eating monkeys, which will be described next.

The Red Colobus is rarely seen in captivity.

Leaf-eating monkeys

The leaf-eating monkeys are confined to the forested areas of the Old World. Colobines are common in India and South East Asia but rare in Africa, where there is only one genus, *Colobus* Their agility in the trees is phenomenal, but they often make a great deal of unnecessary noise. Mature leaves form the basis of their diet but some fruit, buds and flowers are also eaten. They are rarely seen in zoos as it is difficult to give them a suitable diet.

All leaf-eaters have well-separated ischial callosities and, with one exception (the snub-nosed langur), have a long tail. Cheek pouches are absent. The complex digestive system gives

Abyssinian Colobus

them a rather protuberant abdomen. Sexual swelling is not seen in the females at oestrus with the possible exception of the Red and Olive Colobus Monkeys whose habits in the wild are virtually unknown.

Colobus monkeys

Colobus monkeys lack a thumb, and their name is derived from the Greek word 'kolobus' – meaning maimed or mutilated. The muzzle is rather short but the nose is prominent, slightly overhanging the upper lip.

The Red Colobus *(C. badius)* ranges from Senegal to Zanzibar. The guerezas, or black-and-white colobus, have a roughly similar distribution. The King Colobus *(C. polykomos)* has a conspicuous white ruff, and a white tail; one race from Gabon is entirely black. It lacks the long flank fringes of the Abyssinian Colobus *(C. guereza)* which in fact occurs over a wide area in tropical Africa. The Olive Colobus *(C. verus)* from West Africa is relatively small and olive-grey in colour.

Newborn infants are usually a different colour from their parents. Guerezas are white at birth and take about six months to get their adult coloration. The mother allows her infant to be handled by other females and occasionally also by the males.

King Colobus

Leaf-monkeys

The Asian leaf-eating monkeys are found from Kashmir to Ceylon and from China to Borneo. Leaf-monkeys resemble the colobus monkeys but they possess short thumbs. The tail is long and of uniform thickness. The hairs of the crown form a remarkable variety of crests and peaks; the eyebrow hairs are stiff and project forward. Four main groups are recognized.

Firstly there are the Hanuman Langurs *(Presbytis entellus)* which are found throughout India and Ceylon. The montane race of the Himalayas is much larger and heavier than that of the lowlands. It is considered a holy animal and is therefore protected from harm, fed with grain in the temples, and allowed to raid fields and pilfer gardens unmolested. Langurs are preyed upon by Leopards and Tigers so, while they are feeding on the ground, they always remain within a safe distance of trees.

The infant Hanuman Langur is dark brown at birth; soon after its birth it is handed round to all the 'aunties' in the group who are keenly interested in the new arrival.

All the females of the group appear to feel a strong attraction to the dark-coated infant but the males, unlike the colobus males, show no interest at all.

The second group, the purple-faced leaf-monkeys, from southern India and Ceylon, are relatively shy. They are not considered sacred and are shot by the natives for their pelts and for food. John's Langur *(P. johnii)* from southern India is black with long brown hairs on the head; the newborn infant is reddish-brown. Purple-faced Leaf-monkeys *(P. senex)* from Ceylon have grey infants with white cheeks.

The third group of leaf-monkeys (the *'cristata'* group) have bright orange newborn infants while the adults are mainly black, dark grey or ashy. One species, the Silvered Leaf-monkey *(P. cristata)*, has black hairs with silver tips. The Dusky Leaf-monkey *(P. obscura)* from northern Malaya, which has white circles around the eyes, is sometimes called the Spectacled Langur.

This third group of leaf-monkeys, the *'cristata'* group, with

Opposite: Hanuman Langur and infant *(top),* Purple-faced Langur *(centre),* John's Langur *(bottom)*
Below: Dusky Leaf-monkeys with young infant

Golden Langur *(above)*. Capped Leaf-monkey *(below)*

the orange-coloured infants, is found from Bhutan, Assam and southern China, through Burma, Thailand and Vietnam to Malaya, Sumatra, Java and Borneo. There are seven species and many races spread over this vast area. One of the most striking is the Capped Langur *(P. pileata)* from Assam and northern Burma. It is dark grey in colour; the thick mat of dark hairs on the crown contrast with the whiskers and under-parts which are whitish, buff or rusty-red. The most northerly species is the Golden Langur *(P. geei)* from Bhutan. It is cream or golden in colour; the infant is exceptional in being almost pure white at birth.

The fourth group of leaf-monkeys is characterized by having infants which are white with a dark stripe running down the back from head to tail-tip, extending slightly sideways on to the shoulders to form a cross. As the infant grows older, the dark area gradually increases in size until finally only the underparts remain white. In some species all that is left is a white stripe running down the inside of the leg, as seen in the Banded Leaf-monkey *(P. melalophos)* from Borneo. Also from

Borneo, the Maroon Leaf-monkey *(P. rubicunda)* is a rich red-brown with bluish facial skin; the infant is all-white at birth. Other species and races of this group are found in Malaya, Sumatra, Java and Borneo.

Leaf-monkeys live in multi-male groups which vary greatly in size. The male role is that of group-leaders, responsible for directing the movements of the group during feeding and for the selection of sleeping trees. Langurs are mainly arboreal, but Hanuman Langurs spend most of the day on the ground. The females have sole charge of rearing and protecting infants. The strong interest which all the females show in all the infants in their group leads in some instances to a sort of 'baby-sitting' system whereby one female will look after two infants, her own and another's, while the mother forages for food. Sub-adult females are encouraged to handle infants but sometimes their inexperience or awkwardness upsets the infant, provoking screams and wails that prompt the mother to rescue it.

The final group of leaf-monkeys consists of what are some-

Banded Leaf-monkey *(above)*,
Maroon Leaf-monkey *(below)*

Douc Langur *(left),* Snub-nosed Langur *(right)*

times called the 'funny-nosed' leaf-monkeys. All kinds of noses are seen in this group which otherwise shows all the normal characteristics of the subfamily; the specialized digestive system, well separated ischial callosities, no cheek pouches and no sexual swelling.

Douc Langur

The Douc Langur *(Pygathrix nemaeus),* from Vietnam, exhibits the most astonishing coloration of any Old World monkey. It has grey body fur, with bands of black and red at the neck, and pale forearms. The legs are red and the hands and feet are black. The bright colours are emphasized by a large white rump patch and white tail.

Snub-nosed langurs

The snub-nosed langurs *(Rhinopithecus)* are found in the mountains of Szechwan in China. Their beautiful coats consist of milk-chocolate-coloured fur overlaid with long golden strands, and the underparts are orange or yellow. Races from other parts of China have mainly black or dark grey fur with lighter underparts.

Pig-tailed Langur

The Pig-tailed Langur *(Simias concolor)* comes from a small group of islands off the south-west coast of Sumatra, the Mentawai Islands. This dark brown langur is rather maca-que-like in build, being rela-tively short-limbed. The isch-ial callosities are united in the males but separate in females. The snub nose is less devel oped than in *Rhinopithecus*.

Proboscis Monkey

The Proboscis Monkey *(Nas-alis larvatus)* lives in the tropical rain forests and man-grove swamps of Borneo. It is very agile, leaping from tree to tree and even on occasions diving into the flooded creeks of the mangrove swamp. The large males have long, bulb-ous and drooping noses. The female's nose is less developed and slightly tip-tilted; the nose of infants turns up in a most appealing way. The fur is a creamy brick-red which blends into silver-grey on the arms and legs.

Proboscis Monkey groups consist of about twenty ani-mals; there may be several adult males. These warn the group of the approach of danger by giving loud 'honks', the long nose straight-ening out with each 'honk'.

Female Hooloock Gibbon and infant

Apes (Hylobatidae and Pongidae)

The third major group of anthropoids contains the apes and man. The two main external characters that distinguish the apes from monkeys are the lack of any external tail and arms that are much longer than the legs. Only the chimpanzee has sexual swelling at oestrus.

Gibbons and Siamang

Gibbons and Siamangs are considered to be sufficiently distinct from the great apes (chimpanzee, Gorilla and Orang) in their smaller size and their locomotion to be placed in a family of their own, the Hylobatidae. Ischial callosities are present in the gibbons and Siamang, but not in the great apes.

Gibbons are found throughout South East Asia, Sumatra, Java and Borneo. They live in tropical rain forest, deciduous and montane forest. They are wholly arboreal and mainly fruit-eating. Gibbons live in family groups of two to six animals. Males and females are equal in size.

Gibbon locomotion is achieved by arm-swinging beneath the branch; the hands form powerful hooks for suspending the weight of the body.

The coat of gibbons is long, extremely dense and shaggy. The White-handed Gibbon *(Hylobates lar)* from Thailand and Malaya, may be either black or buff with white hands and feet. In the Dark-handed Gibbon *(H. agilis)* from Malaya and Sumatra, the hands and feet match the coat in both black and buff forms. The Silvery Gibbon *(H. moloch)* from Java and Borneo, is grey but very variable in shade. Kloss's Gibbon *(H. klossii)* from the Mentawai Islands is rather small and completely black. In these four species coat colour is not related to sex.

In the two most northerly species, the Hoolock Gibbon *(H. hoolock)* and the Concolor Gibbon *(H. concolor)*, coat colour is sex-related. All infants are pale at birth but soon turn black at about five to seven years. On reaching maturity female Hoolock Gibbons become brown and Concolor females golden or fawn; the male Concolor remains black. In some races Concolor males have white cheek patches.

Concolor Gibbons, female *(left)*
male *(below)*
Male Black Lar Gibbon *(right)*

One of the behavioural characters that distinguish the gibbons and Siamang from the great apes is the very loud calls which they make as part of their regular daily territorial routine. These calls vary with the species and race concerned; in the Lar Gibbon the loudest most remarkable call is that of the adult female, a clear hooting song of rising inflection, soaring high and then falling off in a melancholy cadence. The Siamang *(Symphalangus syndactylus)*, with its extraordinary throat sac, is able to produce two different kinds of note in its song: a deep 'boom' as it sings into the sac with its mouth closed, thus inflating it, and an extremely loud 'wow' made with the mouth open in the ordinary way. These two calls are repeated one after another, at first slowly and then at great speed. The volume of sound is extraordinary.

Siamangs, from Malaya and Sumatra, are slightly larger and more heavily built than gibbons; the arms are even longer, compared to the legs, than those of the gibbon. They are wholly black with some white hairs around the mouth. The coat is less dense than in gibbons and there is some webbing of the toes.

Orang-utan

The great apes (family Pongidae) include the Orang-utan and the chimpanzees and Gorilla. The Orang-utan (*Pongo pygmaeus*) is a single species but there are believed to be two races, one from Sumatra and one from Borneo. Orangs are mainly fruit-eating and live in tropical rain forest. Little is known about their habits but they appear to lead rather solitary lives. Their locomotion is a modified form of arm-swinging; they suspend themselves by their arms, or their legs, or by any combination of arms and legs. The male is much larger than the female and twice as heavy; in adult life he has large cheek pads which in the Bornean race project forwards like blinkers. The male Orang-utan from Sumatra has a flatter face with the cheek pads projecting sideways. The arms are much longer and stronger that the legs; the thumbs and big toes are very short, and the last big toe joint and the big toe nail are missing in the majority of specimens.

Orang-utans from Borneo,
female *(left)* male *(right)*

Chimpanzees playing

Chimpanzees

Chimpanzees live in the tropical rain forests and woodland savannas of Africa from Sierra Leone to Lake Tanganyika in the east. The Common Chimpanzee *(Pan troglodytes)* is found north of the River Congo and the Pygmy Chimpanzee *(Pan paniscus)* south of the Congo, west of the River Lualaba.

Chimpanzees have long, black shaggy coats and young animals have a small white tuft on the rump. They have strong brown ridges, rather flat noses and large protruding ears. A prominent sexual swelling is seen in the females during the oestrous period. Chimps have long arms and rather short legs and, when on the ground, they walk quadrupedally, supporting the front of the body on the back of the knuckles. Pads of hard skin on the middle joints of the fingers make contact with the ground. They live in large multi-male groups which are constantly changing composition. Mothers, their babies and their earlier offspring from the most permanent unit. Chimpanzees have been observed to modify and use twigs as tools. They often insert twigs into termite nests to extract some of the thousands of termites.

Gorillas

Gorillas *(Gorilla gorilla)* are the largest of all primates. They are dignified and reserved in contrast to the excitable and extroverted chimpanzees. In equatorial Africa two main centres of population are found; the lowland gorilla from between the Rivers Niger and Congo and the highland or 'mountain' gorilla from east of the River Lualaba. The lowland race can easily be distinguished by the presence of a slightly overhanging tip to the nose. Gorillas are much larger and heavier than chimpanzees. The male is much larger than the female and twice as heavy. Their coat is thick and mainly black; the mature male has a saddle of grey fur on his back, a 'silver-back male'. The head of the adult male has a conical look due to the bony crests on the skull and the heavy neck muscles which help to support the enormous jaws. Gorillas are almost wholly ground-living and their posture and knuckle-walking gait are similar to that of chimpanzees.

Silver-backed male Gorilla

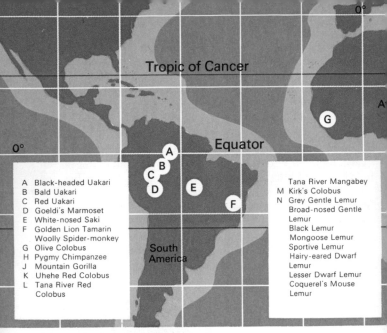

A   Black-headed Uakari
B   Bald Uakari
C   Red Uakari
D   Goeldi's Marmoset
E   White-nosed Saki
F   Golden Lion Tamarin
     Woolly Spider-monkey
G   Olive Colobus
H   Pygmy Chimpanzee
J   Mountain Gorilla
K   Uhehe Red Colobus
L   Tana River Red
     Colobus

     Tana River Mangabey
M   Kirk's Colobus
N   Grey Gentle Lemur
     Broad-nosed Gentle
     Lemur
     Black Lemur
     Mongoose Lemur
     Sportive Lemur
     Hairy-eared Dwarf
     Lemur
     Lesser Dwarf Lemur
     Coquerel's Mouse
     Lemur

Primates in danger of extinction

ENDANGERED SPECIES

Like all wild animals today, non-human primates are threatened with the destruction of their natural habitats. Man's avid hunger for timber, minerals, farmlands and grazing rights is steadily eroding the wildernesses of the world. The International Union for the Conservation of Nature and Natural Resources (I.U.C.N.) in its famous Red Book maintains an up to date list of animals whose continued existence is in danger. The thirty *living* primate species or races listed on this page are liable to become *extinct*, dead as the dodo, in the next decade or so unless action is taken now to conserve the animal resources of the globe by preserving their natural habitats.

As can be seen from the map, Madagascar harbours many species of extraordinary zoological interest, species that evolved in this once-secluded environment, cut off for many millions of years from the rest of the world. The destruction of forests and the introduction of new animal and plant species have altered the ecological balance in such a way that special-

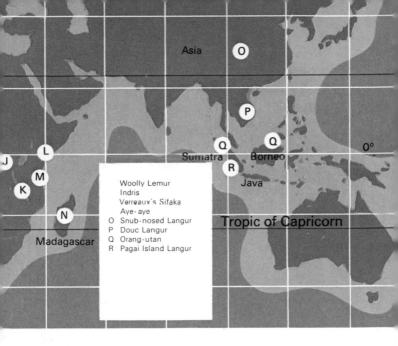

Woolly Lemur
Indris
Verreaux's Sifaka
Aye-aye
O Snub-nosed Langur
P Douc Langur
Q Orang-utan
R Pagai Island Langur

ized primates – like the Aye-aye – probably never very numerous, are now reduced to a few individuals. Efforts are being made to save the Aye-aye by placing several pairs on an uninhabited island off the coast.

At the other end of the scale, those primates most closely related to man, the great apes, are also in grave danger of extinction, the Orang-utan most of all. Wild populations of Orangs in Borneo and Sumatra number less than 5,000 animals. The hunting of Orangs for export has been going on for many years. Usually the mother is shot in order to capture her baby. The population is thus deprived of its most valuable asset from the point of view of conservation – a breeding female. Although the export of Orangs from Borneo is now prohibited, infants can easily be smuggled out. While there is a market for Orangs in the outside world, it is difficult to see how the present low population can be maintained.

It is ironic to think that man, a primate, in failing to protect the vanishing primates, is destroying an important link with his evolutionary past.

PROSIMIAN FAMILIES

Tupaiidae   Lemuridae   Indriidae
Daubentoniidae   Lorisidae   Tarsiidae

Tupaiidae

Tupaia glis	Common Treeshrew
T. splendidula	
T. muelleri	Müller's Treeshrew
T. montana	Mountain Treeshrew
T. javanica	Small Treeshrew
T. nicobarica	Nicobar Treeshrew
T. minor	Pygmy Treeshrew
T. gracilis	Slender Treeshrew
T. picta	Painted Treeshrew
T. palawanensis	Palawan Treeshrew
T. tana	Terrestrial Treeshrew
T. dorsalis	Striped Treeshrew
Anathana ellioti	Madras Treeshrew
Urogale everetti	Philippine Treeshrew
Dendrogale murina	Northern Smooth-tailed Treeshrew
D. melanura	Southern Smooth-tailed Treeshrew
Ptilocercus lowii	Feather-tailed Treeshrew

Lemuridae

Lemur catta	Ring-tailed Lemur
L. variegatus	Ruffed Lemur
L. macaco	Black Lemur
L. mongoz	Mongoose Lemur
L. rubriventer	Red-bellied Lemur
Hapalemur griseus	Grey Gentle Lemur
H. simus	Broad-nosed Gentle Lemur
Lepilemur mustelinus	Sportive Lemur
Cheirogaleus major	Greater Dwarf Lemur
C. medius	Fat-tailed Dwarf Lemur
C. trichotis	Hairy-eared Dwarf Lemur
Phaner furcifer	Fork-marked Dwarf Lemur
Microcebus murinus	Lesser Mouse Lemur
M. coquereli	Coquerel's Mouse Lemur

Indriidae

Indri indri	Indris
Propithecus diadema	Diademed Sifaka
P. verreauxi	Verreaux's Sifaka
Avahi laniger	Woolly Lemur, Avahi

Daubentoniidae

Daubentonia madagascariensis	Aye-aye

Lorisidae

Loris tardigradus	Slender Loris
Nycticebus coucang	Slow Loris
N. pygmaeus	Lesser Slow Loris
Perodicticus potto	Potto
Arctocebus calabarensis	Angwantibo
Galago senegalensis	Bushbaby
G. crassicaudatus	Thick-tailed Galago
G. alleni	Allen's Galago
G. demidovii	Dwarf Galago
G. elegantulus	Needle-nailed Galago
G. inustus	

Tarsiidae

Tarsius spectrum	Spectral Tarsier
T. syrichta	Philippine Tarsier
T. bancanus	Bornean Tarsier, Horsfield's Tarsier

ANTHROPOID FAMILIES

New World monkeys   Old World monkeys   Apes and men

Callitrichidae   Cercopithecidae   Hylobatidae

Cebidae   Pongidae

                                               Hominidae

Callitrichidae

Callithrix jacchus	Common Marmoset
C. humeralifer	Santarem Marmoset
C. argentata	Black-tailed Marmoset
Cebuella pygmaea	Pygmy Marmoset
Saguinus midas	Red-handed Tamarin
S. nigricollis	Black and Red Tamarin
S. graellisi	Rio Napo Tamarin
S. fuscicollis	Saddle-back Tamarin
S. labiatus	Red-bellied Tamarin or White-lipped Tamarin
S. mystax	Moustached Tamarin
S. imperator	Emperor Tamarin
S. bicolor	Pied Tamarin
S. leucopus	White-footed Tamarin
S. inustus	
S. oedipus	Pinché, Cottontop
Leontopithecus rosalia	Golden Lion Tamarin
Callimico goeldii	Goeldi's Marmoset

Cebidae

Pithecia pithecia	White-faced Saki
P. monachus	Monk Saki
Chiropotes satanas	Black Bearded Saki
C. albinasus	White-nosed Bearded Saki
Cacajao melanocephalus	Black-headed Uakari
C. calvus	Bald Uakari
C. rubicundus	Red Uakari
Aotus trivirgatus	Douroucouli
Callicebus personatus	Masked Titi
C. moloch	Dusky Titi

C. torquatus	Widow Monkey
Saimiri sciureus	Common Squirrel Monkey
S. oerstedii	Red-backed Squirrel Monkey
Cebus capucinus	White-throated Capuchin
C. albifrons	Brown Pale-fronted Capuchin
C. nigrivittatus	Weeper Capuchin
C. apella	Black-capped Capuchin
Alouatta belzebul	Red-handed Howler
A. fusca	Brown Howler
A. villosa	Mantled Howler
A. seniculus	Red Howler
A. caraya	Black Howler
Ateles paniscus	Black Spider Monkey
A. belzebuth	Long-haired Spider Monkey
A. fusciceps	Brown-headed Spider Monkey
A. geoffroyi	Black-handed Spider Monkey
Lagothrix lagothricha	Humboldt's Woolly Monkey
L. flavicauda	Hendee's Woolly Monkey
Brachyteles arachnoides	Woolly Spider-monkey

Cercopithecidae

Cercopithecus diana	Diana or Roloway Monkey
C. aethiops	Vervet, Grivet or Green Monkey
C. cephus	Moustached Monkey
C. lhoesti	L'Hoest's Monkey
C. preussi	Preuss' Monkey
C. hamlyni	Hamlyn's Owl-faced Monkey
C. mitis	Blue Monkey, Diademed Guenon
C. albogularis	Sykes' Monkey, White-throated Guenon
C. mona	Mona Monkey
C. campbelli	Campbell's Monkey
C. wolfi	Wolf's Monkey
C. pogonias	Crowned or Golden-bellied Guenon
C. denti	Dent's Monkey
C. neglectus	De Brazza's Monkey
C. nictitans	Greater White-nosed Guenon
C. petaurista	Lesser White-nosed Guenon

C. ascanius	Redtail
C. erythrotis	Red-eared Guenon
C. erythrogaster	Red-bellied Guenon
C. nigroviridis	Allen's Swamp Monkey
C. talapoin	Talapoin or Mangrove Monkey
Erythrocebus patas	Patas or Military Monkey, Red Hussar
Cercocebus atys	Sooty Mangabey
C. torquatus	White-collared Mangabey
C. galeritus	Agile Mangabey
C. albigena	Grey-cheeked Mangabey
C. aterrimus	Black Mangabey
Mandrillus sphinx	Mandrill
M. leucophaeus	Drill
Papio papio	Guinea Baboon
P. anubis	Olive Baboon
P. cynocephalus	Yellow Baboon
P. ursinus	Chacma Baboon
P. hamadryas	Hamadryas Baboon
Theropithecus gelada	Gelada
Macaca sylvanus	Barbary Ape
M. sinica	Toque Macaque
M. radiata	Bonnet Macaque
M. silenus	Lion-tailed Macaque
M. nemestrina	Pig-tailed Macaque
M. fascicularis	Crab-eating Macaque
M. mulatta	Rhesus Macaque
M. assamensis	Assamese Macaque
M. cyclopis	Formosan Rock Macaque
M. arctoides	Stump-tailed Macaque
M. fuscata	Japanese Macaque
M. maurus	Celebes or Moor Macaque
Cynopithecus niger	Celebes Black Ape
Colobus polykomos	King Colobus
C. guereza	Abyssinian Colobus
C. verus	Olive Colobus
C. badius	Red Colobus
C. kirkii	Kirk's Colobus

Presbytis aygula	Sunda Island Leaf-monkey
P. melalophos	Banded Leaf-monkey
P. frontata	White-fronted Leaf-monkey
P. rubicunda	Maroon Leaf-monkey
P. entellus	Hanuman Langur
P. senex	Purple-faced Leaf-monkey
P. johnii	John's Langur
P. cristata	Silvered Leaf-monkey
P. obscura	Dusky Leaf-monkey
P. phayrei	Phayre's Leaf-monkey
P. francoisi	François' Leaf-monkey
P. potenziani	Mentawai Leaf-monkey
P. pileata	Capped Langur
P. geei	Golden Langur
Pygathrix nemaeus	Douc Langur
Rhinopithecus roxellanae	Snub-nosed Langur
R. avunculus	Tonkin Snub-nosed Monkey
Nasalis larvatus	Proboscis Monkey
Simias concolor	Pagai Island Langur

Hylobatidae

Hylobates lar	White-handed Gibbon
H. agilis	Dark-handed Gibbon
H. moloch	Silvery Gibbon
H. hoolock	Hoolock Gibbon
H. concolor	Black Gibbon
H. klossii	Kloss's Gibbon
Symphalangus syndactylus	Siamang

Pongidae

Pongo pygmaeus	Orang-utan
Pan troglodytes	Chimpanzee
P. paniscus	Pygmy Chimpanzee
Gorilla gorilla	Gorilla

Hominidae

Homo sapiens	Man

BOOKS TO READ

The Monkey Kingdom by I. T. Sanderson. Hamish Hamilton, London, 1957.

Primates by S. Eimerl and I. De Vore. Life Nature Library, Time-Life International, New York.

The Life of Primates by A. H. Schultz. Weidenfeld and Nicolson, London, 1969.

A Handbook of Living Primates by J. R. Napier and P. H. Napier. Academic Press, London, 1967.

The Apes by V. Reynolds. E. P. Dutton, New York, 1967.

History of the Primates by W. E. Le Gros Clark. British Museum (Natural History) 8th edition, London, 1962.

Mankind in the Making by W. Howells. Penguin, London, 1967.

My Friends, the Wild Chimpanzees by J. van Lawick-Godall. National Geographic Society, 1967.

The Year of the Gorilla by G. B. Schaller. University of Chicago Press, Chicago, 1963.

Lemur Behaviour by Alison Jolly. University of Chicago Press, Chicago, 1967.

Men and Apes by Ramona and Desmond Morris. Hutchinson & Co., London, 1966.

Primate Behaviour edited by I. De. Vore. Holt, Rinehart & Winston, New York, 1965.

PLACES TO VISIT

GREAT BRITAIN:
Bristol Zoo, Clifton, Bristol, 8.
Chester Zoo, Upton, Chester.
Jersey Zoo, Jersey, C.I.
London Zoo, Regent's Park, London, N.W.1.
Paignton Zoo, Paignton, Devon.
Twycross Zoo Park, Atherstone, Warwickshire.

AUSTRALIA
Adelaide Zoo, Adelaide.
Melbourne Zoo, Melbourne.
Perth Zoo, Perth.
Taronga Zoo, Sydney.

EUROPE
Amsterdam Zoo, Amsterdam.
Antwerp Zoo, Antwerpen.
Basel Zoo, Basel, Switzerland.
Frankfurt Zoo, Frankfurt am Main 1, Germany.
Cologne Zoo, Germany.

UNITED STATES OF AMERICA
Bronx Zoo, New York.
Brookfield Zoo, Chicago.
Goulds Monkey Jungle, Miami.
National Zoological Park, Washington, D.C.
Philadelphia Zoo, Philadelphia.
San Diego Zoo, San Diego.

INDEX

Page numbers in bold type
refer to illustrations.

SOME OTHER TITLES IN THIS SERIES

Natural History

The Animal Kingdom
Animals of Australia & New Zealand
Animals of Southern Asia
Bird Behaviour
Birds of Prey

Evolution of Life
Fishes of the World
Fossil Man
A Guide to the Seashore

Gardening

Chrysanthemums

Garden Flowers

Popular Science

Astronomy
Atomic Energy
Computers at Work

The Earth
Electricity
Electronics

Arts

Architecture

Jewellery

General Information

Arms and Armour
Coins
Flags

Guns
Military Uniforms
Rockets and Missiles

Domestic Animals & Pets

Budgerigars
Cats

Dog Care
Dogs

Domestic Science

Flower Arranging

History & Mythology

Archaeology
Discovery of
   Africa
   Australia
   Japan

Discovery of
   North America
   South America
   The American West